The Whaling Question

(The Inquiry by Sir Sydney Frost of Australia)

Friends of the Earth, San Francisco, and
   The Whale Coalition:
   Animal Protection Institute • General Whale • Greenpeace •
   Oregonians Cooperating to Protect Whales • The Whale Center

Cover, Foreword, and photos copyright   1979
by Friends of the Earth, all rights reserved

Library of Congress Card Catalog Number: 79-2491
ISBN: 0-913890-33-2

Cover art by Larry Foster
Cover design by Marianne Ackerman
Photos by William R. Curtsinger from <u>Wake</u> <u>of</u> <u>the</u> <u>Whale</u> (1979)

Trade sales and distribution by Friends of the Earth
             124 Spear Street
              San Fransisco, CA   94105

Participating members of the Whale Coalition:

Animal Protection Institute	General Whale
P.O. Box 22505	P.O. Box Whales
Sacramento, CA 95822	Alameda, CA 94501
The Whale Center	Greenpeace
3929 Piedmont Avenue	Fort Mason, building 240
Oakland, CA 94611	San Fransisco, CA 94123

Oregonians Cooperating to Protect Whales
873 Willamette
Eugene, OR   97401

FOREWORD

It's not that often that a government changes a firm, longstanding policy overnight. It's even less frequent that such a dramatic and abrupt change comes as the result of citizens' pressure. But this is what happened to Australia's whaling policy in early 1979: Australia switched from whaling nation to leader in the effort to end commercial whaling world-wide. Much of the credit for that astonishing change must go to Australian public-interest organizations and to scientists from around the world.

In the mid-sixties, Australian conservation groups began to focus public attention on whaling. Australia was then one of the few nations that still had commercial whaling stations. Feeling the heat, the Right Honorable Malcolm Fraser, Prime Minister of Australia, called for an independent inquiry into whales and whaling in November of 1977. Several months later the Prime Minister appointed Sir Sydney Frost, a distinguished jurist, as Chairman of the Inquiry. The Inquiry began its task by soliciting testimony from more than 100 individuals and organizations throughout the world. Australian organizations made submissions as well, and the Inquiry commissioned three papers by leading scientists in cetacean biology.

In December of 1978, after reviewing the written evidence and listening to testimony presented at public hearings held throughout Australia, the Inquiry published its summary report and recommendations. The worldwide conservation community has hailed this Report as the most significant development in the effort to save the whales since the 1972 United Nations Conference on the Human Environment voted unanimously for a ten-year moratorium on all commercial whaling. (Japan, the USSR, and Australia, among others, had abstained from that vote.)

The enthusiasm that greets this report stems not only from its laudable recommendations, that Australia prohibit all whaling within 200 miles of its shores, cease the importation of whale products, and seek a worldwide ban on commercial whaling. Conservationists also recognize the Report as the most comprehensive analysis of whaling to have been conducted in many years.

The authority of the Report was ensured by the unquestionable integrity and thoroughness with which the Inquiry considered the issues. Whether discussing cetacean intelligence, the need for whale products and availability of substitutes, the basis for current scientific advice to the International Whaling Commission (IWC), or the hunt itself, the Inquiry considered all perspectives. In every case the Inquiry's conclusion was clear: there are no compelling reasons to continue whaling. On the contrary, the Inquiry found that the evidence pointed to a necessary end to commercial whaling.

Since the Inquiry, evidence that commercial whaling should be banned has continued to mount; the worldwide save-the-whale campaign has continued to find scientific support. At their special meeting in December 1978, members of the Scientific Committee of the IWC studied evidence on population levels of sperm whales in the North Pacific. The Committee startled all attendees to the proceedings by concluding that any level of hunting the Committee might suggest to the full Commission would be scientifically indefensible.

Believing the Inquiry's summary report to be a document of profound importance, we are making it available to all who are concerned for the survival of whales. We are confident that the scientist, the policymaker, and the concerned citizen of the world will benefit from a careful reading of the Report's straightforward review of whaling. The Report presents a stinging indictment of commercial whaling and serves as a brief for putting an end to it. It provides an example worth following for those countries that have yet to call back their fleets. The Report conclusively demonstrates the institutional arrogance of an industry that continues to jeopardize, for the benefit of a few, a heritage belonging to all people.

We wish to thank the conservation community in Australia for its diligence, without which the Inquiry would never have been commissioned or carried to a conclusion of such consequence. We applaud the work of Sir Sydney Frost, whose integrity and dedication is evident in the Inquiry's summary report. Finally, we wish to extend our appreciation to Prime Minister Fraser for his courage and candor not only in commissioning an inquiry into one of Australia's own industries but also in making all the Inquiry's reccomendations official policy of Australia.

                                        Friends of the Earth
Animal Protection Institute
General Whale
Greenpeace
Oregonians Cooperating to
    Protect Whales
The Whale Center

June 1979

INQUIRY INTO WHALES AND WHALING

Chairman: **The Hon. Sir Sydney Frost**
Ph 062 72 5710

Secretary: **Mr A. Struik**
Ph 062 72 3588

P.O. Box E264
Canberra, A.C.T. 2600
Australia

1 December 1978

Dear Prime Minister,

    I have the honour to present with this letter the Report of the Inquiry into Whales and Whaling which you appointed on 20 March 1978.

    As I have noted in the Preface, I was fortunate to receive comprehensive and helpful submissions from many people and organisations interested from various points of view in the conservation of the whale, and the able assistance of the Secretariat provided for me.

Yours sincerely,

Sydney Frost

(Sydney Frost)
<u>Chairman</u>

The Rt Hon. Malcolm Fraser, CH, MP,
Prime Minister,
Parliament House,
<u>CANBERRA</u>   ACT   2600

CONTENTS

Introduction.. 3
1. Whales and their biology............................. 8
2. History of whaling................................... 24
3. Development of control of whaling.................... 38
4. The basis of scientific advice....................... 63
5. International management of whales................... 91
6. Whale stocks and whale research......................102
7. Whale products and alternatives......................126
8. The closure of Cheynes Beach.........................140
9. Whale brains, anatomy and behaviour..................147
10. Techniques used to kill whales.......................172
11. Community attitudes to whaling.......................183
12. Australia's future policy on whaling.................195
Conclusions and recommendations..........................206

Appendixes
1. People and organisations who provided written evidence to the Inquiry as submissions or through correspondence......................212
2. Experts consulted during overseas visit.........219
3. Witnesses at public hearings....................223
4. Press release of the Cheynes Beach Whaling Co (1963) Pty Ltd, 31 July 1978...................226
5. Classification of the order Cetacea.............228
6. International Whaling Commission Divisions and Areas in the southern hemisphere...............233
7. Australian whaling legislation and licences.....234
8. The sperm whale model...........................254
9. Possible objectives for management of marine mammals..263
10. Whaling statistics..............................265

11. An assessment of the sperm whale stock subject
    to Western Australian catching.................270
12. Australian humpback whaling.....................276
13. Summary of products obtained from whales........288
14. Jojoba oil as an alternative to sperm oil.......295
15. Correspondence on the market for sperm oil......306
16. Whale brain and its intelligence potential......317
17. Australian Conservation Foundation opinion poll
    on whaling.....................................325

Bibliography..327

PREFACE

The appointment of the independent Inquiry into Whales and Whaling has its origin in a statement made by the Prime Minister, Mr Malcolm Fraser, in November 1977. The circumstances which led to the Government's decision that an inquiry should be held were mentioned by the Prime Minister in a further statement made on 20 March 1978. The Prime Minister said:

> Many thousands of Australians - and men, women and children throughout the world - have long felt deep concern about the activities of whalers.
>
> There is a natural community disquiet about any activity that threatens the extinction of any animal species.
>
> I abhor any such activity - particularly when it is directed against a species as special and intelligent as the whale.
>
> There are however two distinct views in relation to the activities of whalers.
>
> One view put to me strongly is that all whale species under threat of extinction are protected by moratoriums imposed by the International Whaling Commission and that current policy is in line with the best principles of conservation.
>
> An alternative view which has also been strongly argued to me is that the present practice of killing whales does endanger the whale species. Many other arguments have also been put on both sides.
>
> The Government believed the only fair way to resolve this issue was for an independent inquiry to look at every aspect of whaling ...

The terms of reference for the Inquiry are as follows:

1. The Inquiry shall examine and report upon Australia's policies on whales and whaling. It shall make recommendations on the best way in which Australia might pursue its policy of preservation and conservation of the many species of whales (also known as cetacea).

2. The Inquiry shall in particular examine:
   (a) whether Australian whaling should continue or cease;

(b) the consequences for international whaling of Australia's decision;

(c) international regulatory mechanisms in so far as these bear on Australia's concern for preservation and conservation of whales.

3. Without limiting its scope in any way, in considering Australia's possible actions the Inquiry shall examine:

(a) the role of whales in marine ecosystems and the impact of past and current whale harvesting strategies on the marine environment;

(b) any special features of whales which may make their conservation important;

(c) significant consequences, if any, for other areas of conservation policy;

(d) methods used in taking whales and whether better methods are possible;

(e) factors influencing the scale of Australian and world whaling activity, including the demand for products derived from whales and the possibilities for substitution;

(f) any consequences for Australian employment and industrial development, particularly in Albany;

(g) the implications of Australia's policies on 200-nautical-mile fishing and economic zones in Australian waters, including those adjacent to the Australian Antarctic Territory;

(h) foreign relations aspects;

(i) any other considerations relevant to whales and whaling.

Submissions were sought from many sources. In addition to press advertisements, informal discussions were held with many interested parties and experts. To ensure that the necessary scientific material was obtained, the Inquiry also sent letters to more than two hundred people and organisations in Australia and overseas inviting submissions, and commissioned technical papers related to the terms of reference. The commissioned papers, which have been printed in a separate volume, were prepared by Dr Sidney Holt, Advisor on Marine Affairs, Food and Agriculture Organisation of the United Nations, Rome, Italy; Dr Peter Morgane, Senior Scientist,

Worcester Foundation for Experimental Biology, Shrewsbury, Massachusetts, USA; and Professor Harry J. Jerison, Professor of Psychiatry, University of California, Los Angeles, USA.

The response was gratifying and over one hundred major submissions were received from government authorities, conservation groups, whaling interests, scientists, academics, community bodies and members of the public. Many came from abroad, especially from the United States. All submissions are listed in Appendix 1. In addition many hundreds of people wrote to the Inquiry expressing their views.

In June 1978, I attended meetings in the United Kingdom of the Scientific Committee of the International Whaling Commission and of the Commission itself, and during July 1978, I attended an international conference in Denmark to consider a revision of the International Convention for the Regulation of Whaling. I was accompanied by Mr A. Struik and Mr A. Caton. These meetings were of great assistance to the Inquiry. As well as having informal discussions with delegates and scientists from many nations there was an opportunity to consult many experts both in Europe and the United States. Appendix 2 lists all of the people I had discussions with during this overseas trip. The summaries of discussion at these consultations have also been treated as submissions.

In order to encourage the widest possible discussion of the issues, people and organisations who provided submissions were able to obtain copies of other submissions and commissioned papers on related topics. They could also seek clarification upon statements made in other submissions.

Public hearings were held at Albany, Perth, Sydney and Melbourne. People and organisations who had forwarded submissions were given every encouragement to provide further oral evidence.

Those who came forward to give evidence could be questioned throughout the hearings by representatives of the Commonwealth Government, the Western Australian Government, Cheynes Beach Whaling Company (1963) Pty Ltd, Project Jonah, the Australian Conservation Foundation, Friends of the

Earth and the Secretary of the Inquiry. Arrangements were also made for other people to ask questions either directly or through one of these parties whenever necessary.

It will be noted that the Inquiry had no legal power to compel attendance or administer an oath or affirmation, and absolute immunity did not extend to its proceedings. However, in the case of this Inquiry, its limited powers did not prove to be any disadvantage. Questioning went smoothly and I do not recall any case in which there was a refusal to answer a relevant question.

Originally two major hearings were planned, the first in Albany and the other in Melbourne. The Albany hearing was planned to deal principally with matters concerning the whaling company and the Albany town and region in which the whaling operation is based, and, in particular, with sections 3(e) and (f) of the terms of reference. However, as will appear, this hearing was deprived of much of its substance by the announcement on the first day of proceedings that the whaling company would cease operations in the near future.

The Melbourne hearing was devoted principally to the scientific and foreign relations aspects of the terms of reference, including section 3(g) dealing with 200-nautical-mile fishing and economic zones.

A number of Australian and overseas scientists were invited to attend the Melbourne hearing to discuss their views. Particular mention must be made of Dr K.R. Allen, Mr J.L. Bannister and Dr G.P. Kirkwood from Australia, and Dr S. Holt, Dr M.S. Jacobs and Mr M. Greenwood who travelled to Australia for the Inquiry. These scientists gave extensive evidence, often over several days, which was of the greatest assistance to the Inquiry. A list of all those who provided evidence at public hearings is at Appendix 3.

I wish to record my appreciation of the assistance I received from Mr J.W. Saleeba, Executive Director of Cheynes Beach Holding Limited and Mr G.M. Reilly, a Director of the Company and also General Manager in charge of all whaling operations in Albany, in making arrangements for members of the Secretariat and myself to view the Company's whaling operations. I made flights on the Company's spotter plane, accompanied by Mr Struik, and several voyages on board a catcher. From the plane and on the final voyage I was able to see whales killed. I also visited the station at a time when whales were being flensed (stripped of blubber). All this gave me a clear understanding of the Company's techniques and factory operations.

In conclusion, I wish to acknowledge the valuable support provided by the Secretariat, headed by Mr Andrew Struik as Secretary. The other members of the Secretariat were Mr Brent Arnott, Mr Albert Caton, Miss Michele Foster, Mr Richard Longmore, Mr Peter Rogers and Ms Elizabeth St Clair Long, and administrative and secretarial assistance was provided by Mr Graham O'Neill and Mrs Janis McCulloch. Assistance with final typing of the report was also provided by Ms Wilma Hopkins.

Sydney Frost
Chairman
Inquiry into Whales and Whaling

INTRODUCTION

The Inquiry into Whales and Whaling was established in March 1978 to examine and report upon Australia's policies on whales and whaling including the position Australia adopts internationally, and to make recommendations accordingly. The Inquiry was asked in particular to examine:
  (a) whether Australian whaling should continue or cease;
  (b) the consequences for international whaling of Australia's decision;
  (c) international regulatory mechanisms in so far as these bear on Australia's concern for preservation and conservation of whales.

The full terms of reference are included in the preface to this report.

Scope of the inquiry

It will be noted that the terms of reference extend to 'the many species of whales (also known as cetacea)'. In fact there are over eighty species of the order of mammals called Cetacea, which includes whales, dolphins and porpoises. However, it is only three of the great whales, the right whale, the humpback and the sperm whale, which have been hunted off the coast of Australia, and accordingly this report is concerned mainly with those species.

Two major events occurring after the Inquiry was established have affected its course. The first of these was the announcement on 31 July 1978 by Australia's only whaling company, Cheynes Beach Whaling Company (1963) Pty Ltd (Cheynes Beach), that the 'directors believed operations this year would result in a substantial loss and it was unlikely that there would be any profit in whaling in 1979' and that therefore the 'board had decided whaling operations must end in the near future' (Cheynes Beach Press Release 31 July 1978, see Appendix 4). Cheynes Beach has since announced that it will continue whaling to the end of the 1978 season and will then stop permanently (Cheynes Beach Half Yearly Report, September 1978).

This matter is elaborated in Chapter 8 when dealing with markets for whale products.

The second event affecting the Inquiry was the presentation by Dr K.R. Allen, Mr J.L. Bannister and Dr G.P. Kirkwood of scientific evidence that the sperm whale stock hunted by Cheynes Beach (known as the East Indian Ocean or Division 5 stock) is more depleted than was previously thought. This matter was put before the Inquiry on 21 August 1978 at the Melbourne hearing, with the conclusion that on present analysis males in this stock should be classified as a Protection Stock under the International Whaling Commission management procedure (so that no males can be taken), and that there seemed good reason also to stop taking females immediately, thus helping the stock to recover. Further detail is provided in the discussion in Chapter 6 of the status of whale stocks in Australian waters.

The Inquiry has consequently been relieved of the heavy responsibility of advising upon the future of the Cheynes Beach whaling operation, and accordingly of assessing the economic significance of this operation to Albany.

This does not mean that the Inquiry has no further purpose at all. The question remains of the attitude and course of action which the Australian Government should take on the quota fixed for the East Indian Ocean sperm whale stock by the International Whaling Commission at its meeting in June 1978. This is dealt with in Chapter 6. There is also the possibility, even though remote, that the Australian Government may be faced in the foreseeable future with an application for the grant of a whaling licence. This must be considered - although the state of whale stocks off Australia is in itself sufficient to preclude any such application being granted in the immediate future.

The question of Australia's policy on whales and whaling is also of importance internationally. It is raised because Australia is a member of the International Whaling Commission; as will appear, even in the absence of Australian whaling activities, Australia should continue to have a voice in that forum. Its policy on whaling should also be pursued in bodies such as the United Nations Environment Program, and in Antarctic Treaty negotiations.

Accordingly the Inquiry is not relieved of its responsibility to examine the matters set out in the terms of reference with a view to reaching conclusions and making recommendations upon Australia's future policy. Many of the terms of reference are unaffected in their importance. In particular the Inquiry must consider international regulatory mechanisms in so far as these bear on Australia's concern for the preservation and conservation of whales, the role of whales in the marine ecosystem and the impact of past and current whale harvesting strategies on the marine environment, the implications of Australia's policies on the 200-nautical-mile fishing and economic zones, and the extent and direction of future Australian research on whales.

Major issues

As a background to a consideration of the issues, Chapter 1 includes a discussion of the biological features of whales, Chapter 2 an outline of the history of whaling, and Chapter 3 an examination of the development and the control of whaling, both under Australian legislation and also by the International Whaling Commission, and the extent of Australia's powers in relation to whaling within its fishing zone.

Turning to the issues, there is an international responsibility to ensure that, at a minimum, all whale species are conserved and not endangered. The current Commonwealth Government policy of supporting the conservation of whales based on the advice of the Scientific Committee of the International Whaling Commission is designed to embody this in principle. There is however substantial debate on whether the Commission's procedures are sufficiently conservative and on the reliability of the scientific analysis. These issues are considered in Chapter 4.

Whether Australia decides on a policy of complete protection of whales or one of conservation and careful management, it must recognise that other countries may have different views. Chapter 5 builds on earlier discussions of international regulation of whaling and examines the question of whether the International Whaling Commission is the best body to be responsible for international conservation and management of whales and whether any changes in the Commission are desirable.

The present condition of whale stocks off Australia, and especially the status of the Division 5 sperm whale stock, is examined in Chapter 6. We also consider here the priorities which Australia should adopt in its future research programs on whales.

It has been argued that, even if whales are considered as a potential resource, they should not be killed as there is no need - substitutes being available for all whale products. Australian use, which is primarily of products derived from sperm oil, and the availability of suitable alternatives, are examined in Chapter 7. Chapter 8 then reviews the factors which have influenced the market for sperm oil in recent years, leading eventually to the decision by Cheynes Beach that it would close down.

Some people go further and believe that whales should not be considered as a resource and should be completely protected. It is often claimed in support of this that whales have a high degree of intelligence or mental ability. Chapter 9 reviews and assesses the significance of available material on the brains and behaviour of whales.

Furthermore, many people believe that whaling inflicts an unacceptably slow and painful death on the whale, with an average time to death of about three to five minutes - much longer than would be tolerated for cattle in a slaughterhouse. Chapter 10 reviews the techniques used to kill whales and the available evidence on extent to which these are inhumane.

The most general argument for protection is that the killing of an animal such as the whale requires much stronger justification than now exists or is likely to exist in the foreseeable future. This is part of a growing community concern about wildlife and the environment. Chapter 11 discusses these current community attitudes and also presents an indication of the strength of public concern for whales as expressed in opinion polls and petitions.

The different arguments in the whaling debate are brought together and assessed in Chapter 12. This chapter presents our views on the desirable direction of Australia's future policy on whaling, and is followed by a summary of the Inquiry's conclusions and recommendations.

1. WHALES AND THEIR BIOLOGY

The whale is a warm blooded, air-breathing mammal. It is not a fish, although it is born and suckled, procreates and dies at sea. It has a similar life-span to the elephant, the largest terrestrial animal, and also to man. Because the whale lives in the ocean where little light penetrates, it does not need keen eyesight. Instead it has developed specialised organs to produce and receive sounds, and it uses these to recognise objects and avoid obstacles by echo-location and in the case of sperm whales and other toothed species, to find its prey. They may also be used for communicating with other whales.

In this chapter we do not aim to produce a scientific treatise on the biology of whales, but to refer to those features necessary as background, relevant to the whaling industry, or basic to an understanding of the special place claimed for the whale in the animal kingdom.[1] Thus the whale's feeding habits are relevant to its survival, its reproductive cycle to the size and change in populations, and the social order to harvesting strategies. The whale also has a large brain, claimed to indicate intelligence. We consider the anatomy of its brain and some characteristics of its behaviour important enough to need more detailed treatment and we have presented this in Chapter 9.

General

There are over 80 existing species of the order of mammals called Cetacea. They are divided into two suborders: the Mysticeti (whalebone or baleen whales which strain food through baleen plates that fringe their mouths) and the Odontoceti (toothed whales which are predators dividing their diet predominantly between fish and squid).

---

(1) Much of the material in this chapter is based on Berzin, 1972; Norris, 1966; Ridgway, 1972; and Walker, 1975.

The largest cetaceans are commonly called great whales, or simply whales. These are the 10 species of baleen whale and a single toothed whale, the sperm whale, and they range from about 9 to 30 metres long when mature. Other whales, such as the killer whale, pilot whale and beaked whales, range from 4 to 9 metres long, while the smaller toothed whales are generally referred to as dolphins or porpoises. Dolphins range from 1.5 to 4 metres long and porpoises are the smallest at not more than about 2 metres.

A detailed classification of whale species together with some very brief descriptions is provided at Appendix 5. This illustrates that the order Cetacea has an impressive number of different species ranging from the blue whale - the largest creature ever to have inhabited the earth, reaching a length of 30 metres and weighing up to 150 tons - to the small porpoises only 1.5 metres long and averaging about 50 kg in weight. Some species are localised, such as the Ganges River dolphin, while others, including most great whales, are highly migratory. There are cetaceans which live only in fresh water while others live almost exclusively in the deep oceans - although the majority spend most of their time in the shallow waters over the continental shelf.

Adaptations to marine existence

The ancestors of today's terrestrial mammals emerged from the sea as amphibians and evolved slowly over many millions of years. Later the ancestors of today's whales, primitive carnivores which prowled the shorelines of Africa in particular, gradually returned to the seas, at least 50 million years ago. They probably first entered the rivers, then the estuaries, then the coastal oceans, spreading over the continental shelf and finally ranging the seas from the Arctic to the Antarctic. At each stage of their return to the sea, these animals developed from what they had been, and their bodies became specialised to suit their new habitats (Warshall, 1974).

In the slow process of evolution, the two great sub-orders of cetaceans which emerged were the Mysticeti and the Odontoceti. A third sub-order, the Archaeoceti, became extinct in the lower Miocene period, some 25 million years ago.

Cetaceans gradually took on a streamlined shape, resulting in the hydrodynamically perfect form of today's species. This fundamental adaptation offers least resistence to the water through which the whale or dolphin travels. Resistance to the water is further reduced by the physiological nature of the skin and its cells; natural secretions allow smoother passage. The specifically adapted form and the buoyancy effects of the water explain the large body sizes which certain of the whales have attained.

Streamlining the earlier land body form into one suited to water also required modification of the limbs. Forelimbs evolved to become flippers, hindlimbs degenerated and totally disappeared. Horizontal, flattened tail flukes developed from folds in the skin, as did the dorsal fin in some species.

As warm-blooded animals, whales must maintain a constant body temperature regardless of the water temperature in which they live. With no fur to insulate their bodies, this has been accomplished by the development of a thick (up to 50 cm) layer of blubber, which is situated between the major blood vessels and the skin and serves to keep the body heat inside the animal.

This layer is also important for maintaining a constant temperature. Within the blubber there are blood vessels which can open and close by processes known as vasodilation and vasoconstriction respectively. If the whale's body becomes too hot, blood can be channelled through opened vessels to the skin surface and thus be cooled by the ocean. Similarly, if the whale is cold, the vessels can be constricted, keeping the blood from the exterior regions. The blubber is then returned to its insulating functions.

The heart and blood vessels of the whale contain a much greater volume of blood than terrestrial mammals of equivalent size. The extra blood increases the amount of oxygen which can be maintained in the body and hastens its transport to the vital organs.

Certain questions about the physiological adaptations of whales for diving and tolerating pressure remain to be answered, but some of the general mechanisms are known. To make diving possible, a cetacean breathes the air out from its lungs. It is the oxygen which is combined with the haemoglobin of the blood and with the myoglobin of the muscles which accounts for 80 to 90 per cent of that used during prolonged diving. In fact the proportion of myoglobin in the muscles of whales is very much greater than that found in land animals. Other adaptations include the shunting of blood through the arteries to maintain normal oxygen supply to the brain but reducing supply to the muscles; the loss can be made up when the whale surfaces. A decreased heartbeat (termed <u>bradycardia</u>) contributes further savings of oxygen, and the respiratory centre in the brain is relatively insensitive to an accumulation of carbon dioxide in the blood and tissues.

The high pressures encountered at great depths are alleviated by not breathing air while under pressure, by non-compressible fluids in the body tissues, and by a flexible rib structure which allows the thorax to collapse as pressure increases.

In the case of the sperm whale special adaptations are required. It searches for its prey at great depths (dead sperm whales have been located in twisted submarine cables at depths of up to 2200 metres) and has been known to stay under water for as long as 90 minutes. A special organ found in its head (the spermaceti organ) is thought to enable the whale to control its buoyancy. This is apparently accomplished by changes in the density of the spermaceti oil as the temperature of the water changes (Clarke, 1970 and 1978(a),(b)and(c)). However, it is by no means agreed among scientists that this is the actual purpose of this organ. While some agree that it is a buoyancy regulator for deep diving others regard it as a chamber for reverberation and sound-focussing in the production of echo-location clicks. Still others regard it as a means of assisting in the evacuation of the lungs before diving and absorbing nitrogen at extreme pressures from the air in the nasal cavities before it enters the bloodstream.

It is important to note that cetaceans use sound rather than sight as their main means of determining location and direction and for recognising and communicating. This is another significant adaptation to the marine environment where penetration by light is restricted and eyesight assumes a much less important role. This is considered in some detail in Chapter 9.

Feeding habits

The division of the order Cetacea into baleen and toothed whales reflects a major difference in feeding behaviour and the kind of food taken. It means that baleen whales play a different part in the marine ecosystem from that of sperm whales or other toothed whales. The more aggressive feeding approach required of the toothed whales has also been suggested as leading to significant differences in the behavioural characteristics of the two sub-orders.

All cetaceans have teeth, at least in the gums, during the foetal period. In the Mysticeti, however, those teeth never emerge from the gums, even after birth, but are replaced instead by baleen, or 'whalebone', triangular horny plates made of keratin - similar in substance to fingernails - which emerge from a ridge on each side of the mouth.

In fact these plates are not bone but a modified mucous membrane. The term 'moustached whale', from which the name Mysticeti is derived, refers to the fringes on the plates which protrude from the gums of the adult baleen whale. The baleen plates, rather like a venetian blind on its side, hang from the roof of the whale's mouth. The row on each side of the mouth may consist of more than 300 individual plates, smooth on the outside but fringed on the inside to strain effectively the water that the whale takes into its mouth in large gulps. In this way the baleen whales are able literally to strain the seas for food.

In general the baleen whales feed mainly in the morning and evening. The principal diet, although it varies with the species and the locality, is zooplankton - a complex of small, drifting animals, mainly shrimps of the family Euphausiidae (commonly referred to as 'krill') and copepods, another type of small crustacean, but also including baby crabs, young jelly-fish, certain sea worms and many other creatures. Shoals of these are concentrated in the upper layers of cold ocean waters in summer and it is this that attracts the baleen whales to the polar waters. Some baleen whales will also take fish such as herring, mackerel and sardine, and Bryde's whale has been reported eating small sharks.

Available evidence indicates that the larger southern hemisphere baleen whales - in particular the blue, fin, sei and humpback whales - feed predominantly on Antarctic krill and, taking little food in warmer waters, rely mainly on stored blubber for their nutritional needs.

The two usual methods of taking food are gulping and skimming. In the first, which is used for example by the blue and humpback whales, the animal takes large gulps of water rich in krill. The water is then expelled from the sides of the mouth leaving the krill caught in the baleen sieves. The large flabby tongue is thought to push the water from the whale's mouth, forcing the food to be swallowed to the back of the mouth or gullet, which is surprisingly small, thus limiting the food available to the whale. In the other method, a whale skimming for food swims through the shoal of zooplankton with its mouth open and its head above water to just behind the nostrils. When a mouthful of food has been filtered from the water by the baleen plates, the whale closes its mouth, dives, and swallows. This method is used by right whales and, on occasion, by sei whales.

The feeding habits and preferences of baleen whales are related to the formation of the baleen plates and characteristics of the head, mouth and tongue. The rorquals[1] in particular, with their series of grooves along the throat, are able to increase the capacity for food-rich water by stretching open the grooves like an accordion. It has been estimated that the blue whale, with a strainer (baleen plates) of a total area of 4.5 to 6 square metres, is able to hold over a ton of krill in its stomach and intestine at any one time (Warshall, 1974).

The sperm whale, toothed and without baleen, has a completely different feeding technique, and prefers different food. It feeds mostly on squid, both the smaller species and the giant squids, but also takes deepwater cuttlefish, fish and octopus. Squid bills found in the stomachs of sperm whales are often as large as a child's head. Parts of giant squid estimated

---

(1) The rorquals are the blue, fin, sei, Bryde's, minke and humpback whales. See Appendix 5.

to be as long as 10 metres have also been recovered from stomach contents. Scar marks of 5 to 6 cms in diameter have been noted near the mouth or on the head of sperm whales and it is believed that these are caused by the suckers of the giant squid in its struggles with the whale.

Sperm whales of different ages and sexes prefer different foods. Adult males, which migrate further south than other sperm whales, take squid species that can tolerate a wide range of temperatures and are commonly found in the ocean depths, whereas juveniles and adult females which remain generally in shallower, and warmer, water live on the squid found at various levels in those waters. These differences are important not only in developing policies on whales and whaling, but also when considering the marine ecosystem more broadly.

Whale stocks

Scientists have established that the total population of each of the whale species is not evenly spread throughout the oceans with individuals of the species constantly intermingling. Hence each species cannot be regarded as a single population. Rather, several separate populations, or stocks, between which there is little interchange of animals need to be recognised for each of the different species. Movements between hemispheres are especially rare since the half-year difference in seasons means that the whales in the two hemispheres move towards the equator at different times in following their usual seasonal migrations.

The existence of several stocks of each of the whale species is an important discovery with significant implications for the management (both protection and exploitation) of these species.

The humpback whale provides one of the best examples of segregated stocks. There are reliable data on this species because it remains close to shore and can be easily observed and marked. Based on the evidence of cetologists (Mackintosh, 1942; Chittleborough, 1959; Dawbin, 1959 and 1966), the southern hemisphere population of this species is usually divided into six stocks. Because so few marked whales transfer from one group to another it is apparent that the groups are separate populations with at most a small rate of interchange of individual whales. In particular, Dawbin (1966) has shown through the recapture of marked humpback whales that stocks that breed off Western Australia and those that breed off the east are highly segregated. The stocks off Australia are discussed in more detail in Chapter 6.

The areas of the southern hemisphere in which the different humpback whale stocks are found are assumed also broadly to define separate stocks for other large baleen whales, although the evidence here is much more tenuous. This has led the International Whaling Commission to adopt the following southern hemisphere stock areas for the management of baleen whales:

Area I   -   120°W-60°W
Area II   -   60°W-0°
Area III   -   0°-70°E
Area IV   -   70°E-130°E
Area V   -   130°E-170°W
Area VI   -   170°W-120°W

These Areas, and the sperm whale Divisions described below, are illustrated by the maps in Appendix 6.

Stock divisions in the northern hemisphere are less regular. A description of these can be found in the International Whaling Commission Scientific Committee Reports (see IWC, 1977 and 1978) and the proceedings of the Bergen Scientific Consultation on Marine Mammals (FAO/ACMRR, 1978).

Sperm whales similarly appear to divide into segregated stocks. Some analyses of blood carried out by Japanese scientists indicate that there are two or three stocks in the northwestern part of the north Pacific Ocean (Berzin, 1972). Berzin also refers to research conducted by several cetologists which clearly indicates physical differences between sperm whale populations from different areas, for example differences in the position of the dorsal fin, in the size of the snout protuberance in young animals, and in the length of the tail region, depending upon what region or ocean the specimen came from.

Best (1974) suggested seven or eight southern hemisphere stocks of sperm whales, based principally on density distributions and a consideration of geographical barriers and migration routes. It is now believed that there may be as many as nine stocks, and the International Whaling Commission has reflected this by adopting the following Divisions for management purposes:

Division 1 - West Atlantic (60°W-30°W)
Division 2 - East Atlantic (30°W-20°E)
Division 3 - West Indian (20°E-60°E)
Division 4 - Central Indian (60°E-90°E)
Division 5 - East Indian (90°E-130°E)
Division 6 - East Australian (130°E-160°E)
Division 7 - New Zealand (160°E-170°W)
Division 8 - Central Pacific (170°W-100°W)
Division 9 - East Pacific (100°W-60°W)

The stocks which approach Australia are those in Divisions 5 and 6, with the whales taken by Cheynes Beach being from Division 5. These stocks are discussed further in Chapter 6.

The further identification and assessment of the various whale stocks can be made much easier by marking programs, although there are still many limitations to their practical use. In particular the numbers of whales marked and recovered are relatively small and until suitable external tags are developed whales need to be killed before marks can be recovered.[1]

Migration

Most cetaceans migrate extensively, and almost all species are classified as highly migratory in the latest United Nations negotiating text on the Law of the Sea. Migrations of the great whales in some cases exceed 9000 kilometres each way when travelling to warmer waters in winter to breed and returning to polar waters during the summer months to feed.

---

(1) Best (1976 (a)) reports that a total of 7839 sperm whales had been marked with an internal tag (Discovery-type) up to and including 1974-75, with 4790 being marked in the northern hemisphere. Best comments that the overall recovery rate of marks has been low, and so far it is only from the north Pacific that there have been sufficient mark returns to examine stock identity in any detail. A total of about 20,000 great whales had been marked up to 1973 under various marking schemes. At that time 1192 had been recovered (FAO/ACMRR, 1978).

The best established migration pattern is that of the California gray whale. Earlier overexploited, it is now protected and recovering and provides a tourist attraction off the Californian coast.

These whales spend the summer (June to September) in the northwestern Bering Sea and the Chukchi Sea, and migrate south some 9000 kilometres in autumn and winter to breeding and calving lagoons and estuaries off the coast of Baja California and mainland Mexico. Most of the migrants pass a particular point during a six-week period (from the end of December to the beginning of February) and almost all travel from 3 to 5 kilometres offshore. During the return, northern, migration the whales are more scattered, but they still remain fairly close to the shore line. The whales usually travel singly or in small groups of two or three individuals, at fairly slow speeds.

The humpback and the right whale also migrate close to shore. This is one of the reasons they have previously been heavily exploited, including off the Australian coast. The migration of the right whale is particularly limited; it seems to move only between cold, temperate and sub-tropical waters in both hemispheres, apparently avoiding tropical waters. The migration pattern of the bowhead whale seems to be determined by the seasonal movements of the Arctic ice-edge. Other baleen whales also have a seasonal migration, or at least disperse to warm waters in the winter (although Bryde's whale may remain in warm waters throughout the year).

The rare pygmy right whale has been observed only in the southern hemisphere, mainly in the seas just south of Australia and New Zealand. It is believed that they do not migrate over very long distances, although they move into shallower coastal waters in spring and early summer.

The sperm whale also migrates, travelling between tropical waters in the winter and colder waters in the spring and summer. The range of movements of males and females is different. It seems that females do not migrate into the cold waters but stay within the area between about 40° North and 40° South. This apparent restriction is possibly due to conditions required for breeding and nursing. The adult males wander farther into high latitudes and some may winter in polar and sub-polar waters. Such extensive north-south migration, in the southern hemisphere is supported by recoveries of marked whales. These are consistent also with the longitudinal separation of sperm whale stocks. Although one Soviet mark recovery implies that there may be intermingling across the equator, the fact that the female breeding seasons in the two hemispheres are about six months apart would seem to preclude extensive interbreeding (Best, 1974). It is possible that during the migrations there is some mixing of males from different stocks at high latitudes. In contrast female stocks tend to maintain their integrity.

Sperm whale migration and feeding patterns have been important for the Cheynes Beach operation. Whales approach the edge of the continental shelf and travel westwards, feeding in the deep canyons which occur off this section of the shelf. It is in the neighbourhood of Albany that the continental shelf is narrowest, thus making a land-based whaling operation attractive.

Reproductive cycle

The reproductive cycle and the pattern of social order of whales - including such information as pregnancy rates, gestation and nursing periods, the care of juveniles, and ages to sexual and social maturity - are of fundamental importance in the dynamics of a whale population and hence in assessing the effects on that population of exploitation.

In most cetacean species the period of gestation is from 11 to 16 months. There is usually one offspring, although twins have occasionally been observed with some species. In general the new-born young is one-fourth to one-third the length of the mother.

As part of their adaptation to a fully marine environment, cetaceans have evolved so that both birth and feeding of the young occur in water. Immediately it is born the infant whale must go to the surface to breathe and it is usually helped by the mother or another adult female. The young whale may then suckle with the mother floating on her side to allow the infant to breathe, although later it is suckled underwater. To overcome the feeding difficulties and the limitations imposed by not breathing while suckling underwater, cetaceans have developed body muscles which squirt the milk out of the teats into the infant's mouth. This milk is extraordinarily rich in fat, with only 40 to 50 per cent water compared with 80 to 90 per cent for the milk of most domestic animals. This reduces the time spent suckling as well as keeping to a minimum the amount of water lost to the mother (Slijper, 1962). The rapid growth of most cetaceans (a new-born blue whale, measuring some 7 to 8 metres, almost doubles its length in the first six months of life, and may make a net increase in weight of over 100 kg a day) is related to the richness of the milk.

Depending on the species, lactation continues for 6 to 18 months until the offspring is weaned and ready to fend for itself.

All baleen whales give birth and breed in the warmer waters within their range. In larger species this generally occurs every other year, but in the smaller forms such as the minke whale, most of the mature females give birth once every 12 to 18 months. All baleen whales are monogamous.

The sperm whale differs in many respects from the baleen whales. There are significant differences in size between the sexes, with males growing to lengths of 15 to 18 metres compared with 11 to 12 metres for females. Full size is reached at about 25 years for females and 30 years for males.

Sperm whales are polygynous, with each breeding male having a harem during the breeding season. Mating usually occurs in early summer in temperate waters. The gestation period is up to 16 months with the single offspring (or, very rarely, two) being about 4.5 to 5.0 metres long at birth. Lactation lasts for up to two years and by the end of lactation young may be as large as 6.5 metres. This is generally followed by a resting period, so that a female sperm whale usually gives birth only every four or five years.

At the time they reach sexual maturity, female sperm whales are about 10 years old and 8.5 metres long. The majority of males have reached sexual maturity by the time they are about 20 years old; however, it is not till they are about 25 years old and about 13.5 metres long that they are socially mature and able to participate in breeding.

Age estimates for sperm whales and other toothed whales are based on the number of layers counted in teeth which have been sectioned. For baleen whales, estimates are usually made using the cerumen (wax ear plugs with defined growth rings). While there are some differences of opinion on the interpretation of these data, it is generally accepted that the great whales have a possible life span exceeding 60 years.

Social order

It is recognised by cetologists that the great whales, and indeed all cetaceans, have evolved a social system which has the feature that certain individuals of a species congregate together either continually or at certain times of the year, such as the breeding season, and at other times disperse into smaller groups or as individuals. It follows that if breeding is to occur, some degree of sociability must at certain times be present in every species of whale.

Rorquals are gregarious and are usually seen in groups of a few to several hundred individuals; the larger groups may be scattered over several square kilometres.

The sperm whale is not only gregarious but also polygynous. It has the most developed social order of which the harem is the main feature. Particularly during the breeding season (which peaks in December to January off Australia) the large socially mature males travel with a harem of adult females together with infants, and juveniles of both sexes. The number of individuals in a harem varies from 20 to 40, the average being about 25 (Best, 1978).

Other males live and travel separately from the harems. They tend to travel in groups of 12 to 15 although a number of these groups may travel together. The groups appear to fragment as the members grow older and reach sexual maturity. The adult males that wander farther north and south, some of which may spend the winter months in polar and sub-polar waters, are probably socially mature males which have been unable to gain a harem.

It might appear that males not required for breeding could be taken by the whalers without any collapse of the population. However, additional adult males may be required to stimulate the male with a harem to full breeding activity, or if necessary to provide replacements during the breeding season. It has also been suggested that males with harems may not be replaceable throughout the breeding season because of the earlier return to colder waters of other socially mature males. Scientists have consequently agreed that exploitation should provide for the maintenance of a pool of 'reserve' socially mature males (although the size of the pool is still open to vigorous debate) and that socially mature males should be protected in temperate waters during the breeding season. The International Whaling Commission has accepted these views and declared a closed season for sperm whales over 13.7 metres for the months October to January. This restriction applies north of 40° South in the southern hemisphere, with a corresponding seasonal restriction in the northern hemisphere.

Social order and social behaviour are also of interest in assessing whether cetaceans have a special place in the animal kingdom. This aspect is considered in Chapter 9.

Whales in the ecosystem

Cetaceans live within a rich community of animals and plants and play an important role in the life of the ocean. This community is interconnected in a huge food web or chain.

Baleen whales are at the top of a short food chain starting with phytoplankton (plant plankton) which float in the upper layers of the sea and which synthesise the sun's light into energy. These are in turn eaten by the next link in the chain, the zooplankton, and the zooplankton (particularly krill) are eaten in huge quantities by the baleen whales.

The chain for the sperm whale is a little longer, but functions in the same way. In this case zooplankton provide the food for the fish and squid which in turn are consumed by the whales.

There is thus no direct competition for food between sperm and baleen whales. Since there are several links in the chain between the zooplankton and the large squid hunted by sperm whales, and since sperm and baleen whales mostly feed in different areas, indirect competition is also very limited.

To complete the chain, organic matter - dead animals, waste products from digestion and the like (known as detritus) - sink to the ocean floor and decompose. The organic matter which results is recycled as nutrients for the phytoplankton.

Within each group of animals and plants are those which depend on their ability to exploit 'feasts' of food resources and thereby survive subsequent 'famines' and those which are best adapted to exploit food resources available in a more steady stream (Tranter, 1978). Baleen whales follow this latter pattern; they are large, long-lived animals, well-insulated against the cold and capable of large-scale migrations to exploit favourable environments.

Their fecundity and breeding rate are low, they invest a large part of the resources available to them in parental care including producing milk for their young, and they share the common food source of zooplankton with many smaller animals such as crabeater seals, fur seals, penguins, other birds, squid and fish (Tranter, 1978).

The question arises whether man's rate of harvest of whales permits these slow breeders to replace the whales taken. While there is some evidence of increased pregnancy rates in depleted baleen whale stocks (see, for example, Laws (1962) and Lockyer (1972)), whales are not adapted to respond quickly to environmental fluctuations nor to colonise by rapid breeding environments which favour their survival (Tranter, 1978). Apart from the California gray whale, the evidence for recovery of protected baleen whale stocks is inconclusive.

As the total living mass of whales decreases, more nutrients and energy pass to the remaining organisms in their ocean ecosystem. The result can be an increase in these populations and a generally less stable environment, leading to disruption of the resilience and stability of a finely balanced ecosystem. With the greater abundance of krill the imbalance is compounded by the resultant population increase of animals which feed on krill such as crabeater and fur seals and penguins. The problems faced by whales in their ecosystem are examined in Chapter 4.

2. HISTORY OF WHALING

Although the total history of whaling activities[1] can be measured in some thousands of years it was only with technological developments of the late eighteenth and nineteenth centuries that man's efficiency in hunting whales really increased - in line with his general ability to manage, or destroy, his environment. By the end of the nineteenth century some whale stocks were approaching extinction. The pattern of recognition of whaling potential in a region, establishment of an industry, overexploitation and a final collapse was already well established - and was to be repeated throughout the twentieth century.

While in the last two or three decades scientists have developed greater expertise in advising on the effective management of whale stocks, they continue to be hampered by limited knowledge of whale populations and of the biology and behaviour of whales. The regulation of whaling has also been frustrated by the unwillingness of decision-makers to respond promptly to scientific advice and by the difficulties of reaching international agreements on quotas. Thus, many of the issues of present concern are not new.

Early whaling

Drawings of whales by primitive man have been dated as early as 2200 BC. Subsistence whaling by Alaskan Eskimos and also by Norwegians as a source of meat and for the provision of clothing and implements is estimated to have taken place as early as 1500 BC and 890 AD respectively. Subsistence takes continue to this day, involving for example bowheads in Alaska, gray whales off the Pacific seaboard of the USSR, and humpbacks to a small extent in Tonga and Greenland. This has caused some problems for the

---

(1) Much of this chapter, and especially sections on Australian whaling, is based on the Commonwealth Government Agencies' and Departments' submission to the Inquiry.

International Whaling Commission because the bowheads and the humpbacks are protected, and the gray whale is only recently recovering.

In the eleventh and twelfth centuries, the Basques hunted the black right whale in open boats, using hand harpoons, in the coastal waters of the Bay of Biscay. At first operations were shore-based but as local whale populations decreased, the Basques went wider afield. Whaling activities also spread to other countries in northern and central Europe. With the development of tryworks (furnaces and boilers for extracting oil from blubber), which could be placed on board sailing-ships, whaling was able to be conducted entirely at sea, that is it became 'pelagic'. This was the beginning of an industry which lasted over many centuries.

During the seventeenth and eighteenth centuries the English and Dutch caught right and bowhead whales successively from Spain to Spitzenbergen, Greenland and Hudson Bay. The discovery of sperm whale oil as a superior lamp fuel early in the eighteenth century led to further expansion of the industry. The American fleet based in New England increased rapidly during the late eighteenth and early nineteenth centuries and the Americans replaced the Dutch as the dominant whaling nation.

A British whaler, the _Emilia_, rounded Cape Horn on a whaling expedition in 1788, returning to England in 1790 with a full cargo of sperm oil from the south Pacific. Other American and British ships were soon to follow and a substantial high-seas whaling industry soon developed in the Pacific.

Throughout the eighteenth and nineteenth centuries, whalers - particularly from North America and Britain - concentrated on the slower swimming species, including right, bowhead, gray and sperm whales. The whaling industry provided the main ingredients for perfumery and toiletries as well as cooking oils. Sperm oil was used for lamps, spermaceti wax for candles, and whalebone (baleen) for corsetry and numerous other products.

The application of petroleum and gas in lighting reduced the need for whale oils (especially sperm oil). Declining prices, coupled with growing scarcity of sperm whale and easily caught baleen stocks, resulted in a decline in whaling activity during the last half of the nineteenth century, although very high prices for whalebone kept whalers out hunting for surviving stocks of bowheads in the Bering Sea and gray whales on the west coast of the United States.

Early Australian whaling

Following the development of pelagic whaling in the Pacific Ocean in the latter part of the eighteenth century and Indian Ocean in the early part of the nineteenth, whaling fleets from many nations based operations on ports in New South Wales and Tasmania. There is ample contemporary evidence of the abundance of whales off those shores. In the year 1791 a sea captain wrote to the owners in London - they were the owners of Emilia also - that off the coast of New South Wales close to the latitude of Port Jackson 'within three leagues of the shore we saw sperm whales in great plenty', that he was sailing through different shoals of them throughout the day, and that they were all round the horizon as far as he could see from the mast head (Dakin, 1977, p.10). In 1804 the commanding officer in Hobart Town said, in a letter to Sir Joseph Banks, that the river had for some six weeks been full of right or black whales and that 'three or four ships might have lain at anchor and with these filled all their casks' (Dakin, 1977, p.31-2).

The first Australian-built deep-sea whaler was launched in 1805, and by 1819 Sydney owners held considerable whaling interests. Sperm whalers operated from Sydney Harbour and Twofold Bay in New South Wales. There were nine deep-sea whalers operating from Twofold Bay when operations came to an end in 1848.

There were also many bay whaling stations established in Australia during the early nineteenth century. These shore stations concentrated on the female southern right whale which came into bays or estuaries to give birth.

Bay whaling stations were set up on the Derwent Estuary, and along the coast of Tasmania from Recherche Bay to Falmouth Bay. At the same time, bay whaling stations were established on the Australian mainland in

New South Wales (Twofold Bay), Victoria (Portland and Port Fairy), South Australia (Encounter Bay and near Port Lincoln) and Western Australia.

By the 1830s the colonies supplied most of the equipment used by the whalers. In 1826 the total value of whale exports from New South Wales was £34,850. By 1840 it had jumped to over £335,000 with whales being taken as far afield as New Zealand. The Australian gold rush of the 1850s attracted many tradesmen and whaling crews away from the industry, creating a shortage of men and boats. The value of whale exports from New South Wales fell to £16,000 by 1853, partly due to the effect of the gold rush and partly due to the decline in southern right whale stocks. Whaling from Australia continued to decline and remained at a low level until the twentieth century (Colwell, 1969).

Modern whaling

Modern whaling was a Norwegian development. From 1864 to 1868, Svend Foyn developed a cannon-fired harpoon which could be used on a steam-powered chaser. By 1872, this harpoon had been perfected: an umbrella-shaped, 55 kg steel projectile with an explosive head which fragmented three seconds after penetrating the whale. This enabled whalers to turn their attention to the fast swimming rorquals (the blue, fin, sei and Bryde's whales) which previously could not be caught by open-boat whaling methods.

At first blue and fin whales in the northern Atlantic were hunted but stocks declined quickly. However, the use of steam-powered ships and modern equipment enabled whalers to voyage further and exploit whales from all the world's oceans including the Antarctic regions. Early Antarctic whaling was land-based, principally from South Georgia, with 90 per cent of the catch consisting of the slower moving humpback whales. In the 1910-11 Antarctic season, 8294 humpbacks were killed compared with only 393 blue whales. As humpback whales became scarce, the fleets began hunting the blue whales. The total Antarctic catch in 1913-14 amounted to almost 9500 whales representing 40 per cent of the world catch.

World War I prevented further expansion in whaling activity and it was not until the 1924-25 season that the total world catch of whales amounted to more than 23,000, comparable with pre-war figures.

Cutting up whales alongside the ship often proved impossible in the Antarctic because of gales and ice. In 1925 factory ships were introduced with large slipways at the stern. This allowed crews to haul whales aboard at sea, thus enabling a fully pelagic Antarctic whaling operation independent of land-based stations. World catches of whales increased from approximately 23,000 in 1924-25 to 43,130 in 1930-31 largely through expansion in pelagic whaling.

Whale oil production increased almost eight-fold from 1921 to world record levels in 1931. Oversupply of the oil and falling world prices, coupled with the failure of governments to achieve substantial international agreement, led to independent efforts by whaling companies to limit the production of whale oil in 1932. Individual companies were assigned harvest quotas, on the basis of the amount of oil they were able to place on the market. The cartel of whaling companies devised a blue whale unit to bring oil production from different whale species to a common unit for the assignment of quotas. Thus one blue whale equalled either one blue, three humpback or five sei whales. The cartel broke down by the 1934-35 season as the price of oil failed to respond to reductions in production and quotas became increasingly difficult to negotiate between companies.

During the period between wars there were many attempts to regulate whaling internationally. In 1924 and 1927 there were unsuccessful efforts by the League of Nations to place controls on whaling. At Geneva in 1931, 21 countries (including Australia) signed a limited agreement called the 'Convention for the Regulation of Whaling'. This Convention applied only to baleen whales and was at best a tentative step towards conservation. 'Immature, undersized' whales were granted protection, but the terms were never defined. Right whales were given limited protection. No quotas were established and no attempt was made to reduce the take of whales by shortening the season or placing limitations on the numbers of ships used.

By 1 January 1976, 47 nations had signed this treaty which is technically still in force (Scarff, 1977).

In 1937, following an international whaling conference held in London, representatives of some of the nations engaged in whaling - Argentina, Australia, Germany, Iceland, New Zealand, South Africa, the United States and the United Kingdom - signed an agreement that brought into effect a more substantial convention regulating whaling. This determined a closed season for factory ships in the Antarctic, set minimum size limits for blue, fin, humpback and sperm whales, prohibited the killing of gray, right and bowhead whales and provided for the protection of female whales attended by calves. Rules were also established to ensure the full use of all whales caught and the collection of basic statistical information (Gulland, 1974).

In the 1938-39 season, there were 16 shore stations, 37 factory ships and 362 catcher boats operating in the Antarctic. The factory ships came from seven different nations; thirteen from Norway, ten from England, six from Japan, five from Germany, and one each from Panama, the United States and the Soviet Union. A total catch of 38,356 whales was taken in the Antarctic in the 1938-39 season representing 83.8 per cent of the total world catch.

Pelagic whaling in the Antarctic virtually stopped during World War II. When conferences recommenced in 1944, the blue whale unit was changed to equal one blue whale, two fins, two and a half humpbacks or six sei whales, and a quota of 16,000 units was set for the Antarctic region. The first post-war Antarctic season began in November 1945. Approximately 13,500 whales were taken by nine whaling fleets during the 1945-46 Antarctic season. Following this result which was disappointing to the whalers, and against the background of the post-war spirit of international co-operation, the United States convened an international whaling conference in Washington DC on 20 November 1946. This meeting drafted a convention which all major pelagic whaling nations signed and later ratified as the International Convention for the Regulation of Whaling. This established the International Whaling Commission. Australia was an original signatory to this Convention.

The International Whaling Commission has dictated the pattern of whale management and conservation from the time of its establishment to the present day. The history of the Commission and its management procedures are discussed further in the next three chapters.

In the 1946-47 season, the first reasonably normal post-war season, whaling nations had 20 factory ships and 246 catcher vessels in operation. Four years later, the number of factory ships had increased to 26, while the number of catcher vessels had nearly doubled.

Subsequent reductions in world whale catches through declining whaling stocks and increasing conservation measures by the International Whaling Commission resulted in substantial excess fleet whaling capacity. Sales of pelagic expeditions by various companies had as their main purpose the transfer of quota shares between countries. In some cases after expeditions were sold and quotas transferred to the purchaser, the vessels themselves were not moved. This market enabled Japan, which had a low-cost, efficient industry and also a home market for meat, to increase quota shares, while European countries (the United Kingdom and Netherlands and Norway to a lesser extent) were able to withdraw from whaling without total loss (Gulland, 1974).

Whaling activity, which had reached a peak of 66,090 whales during the 1961-62 season, had fallen by 1975-76 to a total world catch of 19,337 whales taken by 117 catchers servicing eight shore stations and eight pelagic factory ships.

Australia has been the last English-speaking nation to whale commercially. The last of the United Kingdom's fleets was sold in 1963. New Zealand shut down its whaling operations at the same time. The United States stopped whaling in 1972, Canada in 1973 and South Africa in 1976.

A history of overexploitation

The history of whaling, particularly in the twentieth century, has been one of excessive exploitation followed by the economic collapse of whaling operations when whale populations dropped to very low levels, sometimes almost to the point of extinction.

The history of seasonal catches from 1910 to 1976 in terms of major species is shown in Figures 2.1 and 2.2. These figures demonstrate that the industry has been based on each of the major species of whale in turn, starting with the largest and most valuable. As the number of each preferred species declined, attention was directed towards the next largest species - from the blue whale to the fin whale before World War II, to the sei whale around the early 1960s, to the minke whale in the 1970s.

The enforced progression from the larger whales to the smaller species was eased in economic terms by a major change in the use of whales. Originally the only significant product from whales was oil, but since the 1950s more use has been made of meat, which was previously discarded. The main uses of the meat are as extracts or as frozen meat for pet food and also in some countries for human consumption. The value of baleen whales has thus increased considerably.

Catches of sperm whales increased significantly following World War II as the value of sperm oil in applications such as extreme pressure lubrication and industrial waxes became recognised.

Table 2.1 provides estimates for different species of large whales both for recent world population levels and also for levels before exploitation by man. These figures are intended to be illustrative and to give only a general idea of the numbers involved; most are in fact revised regularly as part of the International Whaling Commission's assessment procedures. However, they demonstrate clearly the extent to which whale populations have been depleted.

Modern Australian whaling

Twentieth century whaling by Australia has essentially concentrated on two species, the humpback whale and the sperm whale. As this brief survey shows, the overall pattern of high exploitation followed by the economic collapse of the industry also applies to Australian whaling.

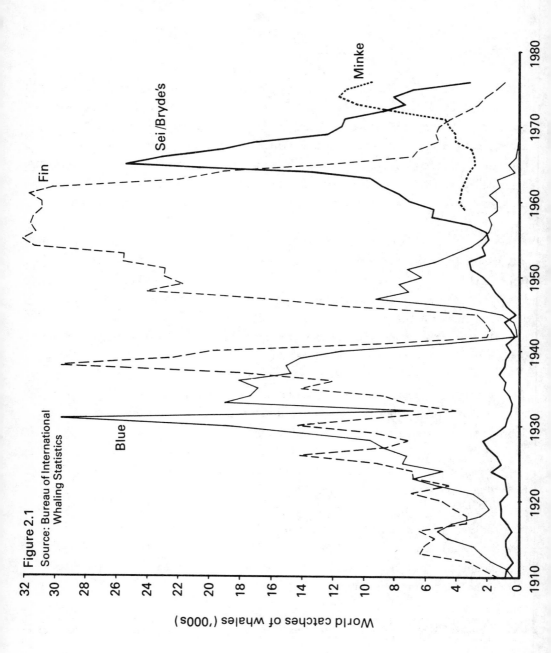

Figure 2.1
Source: Bureau of International Whaling Statistics

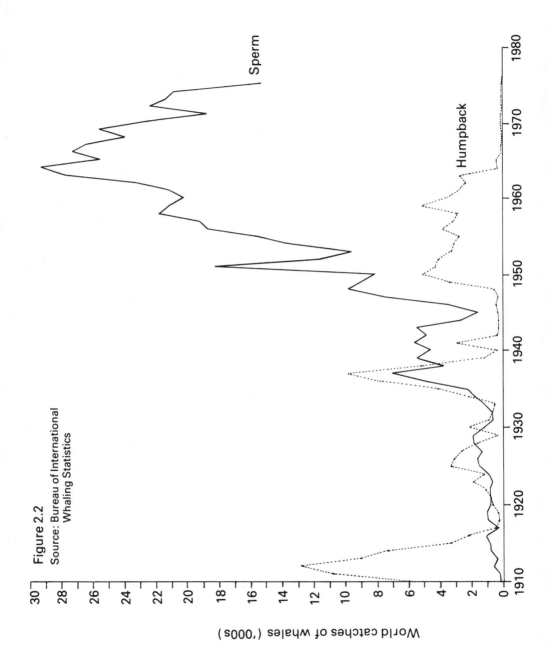

Figure 2.2
Source: Bureau of International Whaling Statistics

Table 2.1
Whale populations[1]

Species	Estimates before commercial whaling ('000)	Recent estimates ('000)	1979 IWC catch limits
Blue	196	10	0
Humpback	100+	5	10*
Fin	448	102	470
Sei (includes Bryde's)	198	74	538
Minke	250+	200+	9173
Right	50[2]	4[2]	0
Bowhead	50 to 56[3]	1.3 to 2.7[3]	27* struck or 18 landed
Sperm	950+	610+	6121 (excludes limits to be set for north Pacific)
Gray	15[2]	11[2]	178*

* Assigned for subsistence take

---

(1) Estimates are of numbers of whales of exploitable size. Unless otherwise indicated, they are taken from Allen and Bannister (1977).
(2) Scheffer (1976).
(3) Report of the Scientific Committee to the 29th Meeting of the International Whaling Commission (see IWC, 1978).

Humpback whales off Australia were hunted intermittently before World War II. Norwegian whalers established a shore station at Jervis Bay (New South Wales) in 1912, taking 537 whales before stopping operations in 1913. Norwegian companies also operated shore stations in Western Australia at Albany (Frenchman's Bay) from 1912 to 1916 and at Point Cloates from 1913 to 1916. The Point Cloates station was operated by an Australian company from 1922 to 1924, when it was taken over by a Norwegian company. The station again closed in 1929 because of a dispute with the Australian Government over tariffs. However, it was not until 1935 to 1939, when there was extensive hunting by foreign fleets off the Western Australian coast and on the Antarctic feeding grounds, that the population declined markedly.

The period between the two World Wars also saw the end of another Australian activity which had continued quietly for nearly 100 years. The bay whaling station at Twofold Bay took its last whale in 1926. As something of an anachronism the whalers there had continued chasing right and humpback whales in open row boats, towing them back to the tryworks on shore. A unique aspect of the catching operations there was the help the whalers got from the presence of a pack of killer whales which 'bailed-up' passing whales and hindered their escape from the whalers.

Australian shore-based humpback whaling was re-established in 1949 when the Nor' west Whaling Company reopened the old station at Point Cloates. At about the same time the Commonwealth Government was seriously considering the operation of a factory ship and attendant catcher boats in the Antarctic. In 1949 it established the Australian Whaling Commission. Interest was, however, transferred to the establishment of land-whaling stations off the west coast of Australia and until 1956 a shore station at Carnarvon, Western Australia was operated by the Commission. Its assets were then sold to the Nor' west Whaling Company.

In 1956 there were six land-based stations operating in Australia; these were at Norfolk Island and at Tangalooma and Byron Bay on the east coast, and at Carnarvon, Point Cloates and Albany on the west coast. However, by 1962 the humpback whale fisheries had collapsed and all whaling operations on humpbacks had ended by 1963.

Dr R.G. Chittleborough has suggested that, if allowed to recover to their original populations (which may take from 35 years on the west coast to 50 years on the east coast), the western and eastern stocks of humpback whales could sustain annual catches of 350 and 300 per year respectively. Catch levels of humpbacks during periods before the collapse of the fishery were, however, substantially higher than these (see Chapter 6 and Appendix 12).

It is informative to note some of the reasons given by Chittleborough for the failure of the humpback fisheries off Australia. He believes that:
Partly because the annual catch was far too high and partly because of a lag in identifying the key research data required, the information essential to determining realistic limits became available too late to be effective. There was also an unwillingness on the part of decision-makers to accept the seriousness of the decline when warned by research workers (Chittleborough, 1978, p.7; see Appendix 12).

During the last century, sperm whales were taken by American whalers close to the continental shelf between Cape Leeuwen and just east of Eyre in the Australian Bight, and between 1912 and 1916 they were taken by the Norwegian company based at Albany. This led Cheynes Beach, which had begun humpback whaling at Albany in 1952, to survey sperm whale stocks in that area. From 1955 to 1963 the company hunted both species, but with sperm whales becoming progressively more important. Since 1964 the operation has been solely based on sperm whales and has been the only active whaling operation in Australia during this period.

The whale stock from which catches are taken at Albany is designated by the International Whaling Commission as the East Indian Ocean stock, or Division 5. Since 1946, approximately 30,000 sperm whales have been taken from this stock, of which a little more than 3000 have been females. Very few females were taken until the mid-1960s, and only in the early 1970s were more than 200-300 taken in any one year. Up to 1956, the average catch of males was less than 500 a year, while from 1960 until 1974 around 1000 a year were taken, with the greatest number (just over 2000) caught in 1965. The largest catches taken by the Albany operation were, for males, 823 in 1971 and, for females 480 in 1975 (Bannister, 1978).

Chapter 6 provides details of the total annual catches taken from the stock, as well as those of the Albany station alone.

It now appears that catches in recent years were substantially above levels that could be sustained, leading to a population significantly more depleted than was earlier believed. In particular, there may not now be sufficient socially mature males in the population to maintain potential reproduction. This has led scientists to take the view that no more whales should be taken from this stock in the immediate future (Kirkwood, Allen and Bannister, 1978). Chapter 6 discusses this matter in greater detail.

3. DEVELOPMENT OF CONTROL OF WHALING

As indicated earlier, international co-operation on the management of whale stocks has been conducted principally through the International Whaling Commission since the Second World War. The development, structure and operation of the Commission are thus relevant to our task, and we provide here a factual account prepared by Dr K.R. Allen at the request of the Inquiry. This provides background for the review of scientific advice and the examination of criticisms of the Commission in later chapters.

Australia has been a member of the International Whaling Commission since its beginning and the various Australian Acts dealing with the control of whaling largely reflect its responsibilities as a party to the International Whaling Convention. However Commonwealth and State legislation dealing more broadly with wildlife can also be used to implement policies on whales. This chapter examines existing legislation and draws attention to the powers available for the management of whales, particularly with the pending declaration of a 200-nautical-mile fishing zone.

The International Whaling Commission

Since 1948 whaling has been regulated internationally by the International Whaling Commission, which was established after a series of conferences between 1944 and 1946. Membership of this Commission has been fairly constant, and has varied much less than the list of countries engaged in whaling at any one time. All but one of the original member countries are still members. They are Argentina, Australia, Brazil, Canada, Denmark, France, Iceland, Japan, Mexico, Netherlands, New Zealand, Norway, Panama, South Africa, USSR, United Kingdom, and USA. Some of these, such as the Netherlands and New Zealand, have resigned for a time and have now rejoined. Of the present 17 member countries, eight - Australia, Brazil, Denmark, Iceland, Japan, Norway, USSR and USA - are still engaged in whaling operations to some extent. In recent years the number of large whales taken by the member nations has averaged about 90 per cent of the total world catch, and has produced about 95 per cent of the oil.

Several countries which are not members of the Commission also carry on coastal whaling operations. These include Peru, Chile, Portugal, Spain, and South Korea. In addition there has been for several years a small pelagic operation using a combined catcher-factory ship. This has been based at various times in different countries.

Table 3.1 lists the countries which are now known to be engaged in whaling, or which are now members of the Commission, and shows their most recent known annual catches.

Table 3.1
Numbers of whales taken in a recent year

Countries whaling in 1978		Number of whales taken[a]
Members of IWC:[b]	Australia	624
	Brazil	1 030
	Denmark	182
	Iceland	580
	Japan	8 999
	Norway	1 772
	USSR	12 139[c]
	USA	29[d]
		25 355 = 89.8 per cent of total
Others:	Chile	62 (76)
	Peru	1 511
	Portugal	238 (75)
	S. Korea	574 (75)
	Spain	224 (74)
	Various[e]	276 (75)
		2 885 = 10.2 per cent of total

(a) Figures are generally for the 1976-77 and 1977 seasons, but where these are not available figures are given for the year indicated.
(b) Members of IWC not whaling in 1978 were Argentina, Canada, France, Mexico, Netherlands, New Zealand, Panama, South Africa, United Kingdom.
(c) Includes aboriginal catch of 186 gray whales.
(d) Aboriginal catch of bowheads.
(e) Independent catcher-factory ship registered in various countries.

Objects of the Commission: While the primary considerations which led to the establishment of the International Whaling Commission sprang from the desire to maintain the industry in a flourishing condition, the agreement on which it was based recognised that it was essential for this purpose to maintain the whale stocks themselves, and that the interests of future generations must be considered. The Convention under which the Commission was established recognised in its preamble 'the interest of the nations of the world in safeguarding for future generations the great natural resources represented by the whale stocks'. It also states the purpose of the Convention as 'to provide for the proper conservation of whale stocks, and thus make possible the orderly development of the whaling industry'.

Types of regulation: The Commission lays down a set of regulations for the management of the whaling industry which are revised annually. These constitute the Schedule to the Convention.

These regulations provide for:
- classification of stocks of whales, and protection of certain species or stocks;
- catch-limits by species, areas, and sometimes by sexes for those whales which may be taken;
- minimum size-limits, and in one case a maximum size-limit;
- duration of whaling seasons;
- limitations on areas where factory ships may be used;
- efficient treatment of carcasses;
- inspection of whaling operations;
- collection of data.

All amendments to the Schedule have to be adopted by a three-quarter majority of those member nations of the Commission who vote on them. Each amendment becomes effective 90 days after notification to member governments unless one or more governments register an objection within that time. If any objections are registered the amendment does not become effective until the end of a further 90-day period from the expiry of the original period. Further objections may be registered during this second 90-day period, or during an additional 30-day period from the date of the last objection received during the second period. At the end of the second 90-day period, or, if appropriate, the additional 30-day period, the amendment becomes binding on all member governments which have not registered an objection. It is not binding on those members which have registered an objection.

While the objection procedure is sometimes adversely criticised as making the Commission ineffective, it is now little used in practice. At the present time there is only one objection in effect; this, which relates to the restriction to eight months of the sperm whale season, is by Australia only and has stood since 1952. Since 1973 no other objections have been registered, and few earlier objections have endured for more than a single year. Thus, the prevailing situation is that provisions adopted by the Commission, even by a narrow margin, are in effect binding on all its members.

Enforcement of regulations: The Schedule provides that all whaling operations shall be subject to inspection by the government concerned, and that all infractions must be reported by the government to the Commission. Countries are also required by the Convention to provide under their own laws penalties for infractions of the Schedule, and operators may not pay bonuses for whales taken in breach of the Schedule. The reported infractions are examined at length by the Commission at each of its meetings.

However, despite these regulations, there is little doubt that some breaches have occurred which have not appeared in national infractions reports.

In 1972 the Commission, after many years of negotiation, strengthened its controls further by introducing the International Observer Scheme. Under this scheme most whaling operations controlled by the Commission are observed on its behalf by representatives drawn from a different member nation. These observers provide an additional check to the supervision by the government inspectors.

Technical Committee: The Technical Committee was originally intended to review and report to the Commission on technical matters affecting its operations. For a time it functioned in this way as a closed Committee. However it has for many years now functioned as a less formal meeting of the Commission itself, at which all matters relating to the technical, and particularly the regulatory, aspects of the Commission's work are discussed, national attitudes brought out, and preliminary decisions reached on a simple majority basis. Recently the need for a separate technical forum has been felt, particularly to examine the technical, operational and social aspects of the Scientific Committee's recommendations. In 1978 therefore the Commission decided that the Technical Committee should be reconstituted and should meet separately in advance of the Commission to consider, in particular, management principles taking into account the recommendations of the Scientific Committee, and technical and practical options for their implementation.

Scientific advice: From its early days the Commission recognised that it would need scientific advice on the state of the whale stocks and how they should be regulated. It therefore set up an ad hoc Scientific Committee, consisting of scientists appointed by the national Commissioners for this purpose. At first this Committee functioned somewhat irregularly, and sometimes had difficulty in reaching firm conclusions on the advice that it should give to the Commission. About 1960 the Commission recognised that it could benefit from an independent review of the situation. It therefore set up the Committee of Three Scientists, which consisted of three scientists who had experience in population dynamics, and were drawn from countries not engaged in pelagic whaling. This Committee, later extended to four scientists, was asked to deal specifically with the baleen whales of the southern hemisphere. It worked from 1961 to 1964 in close co-operation with the Scientific Committee of the Commission, and provided firm advice on the state of these stocks, and the consequences to them of various actions which the Commission might take. After 1964 the Food and Agriculture Organisation of the United Nations (FAO) carried out stock assessments on behalf of the Commission until 1969.

From 1961 onwards the Scientific Committee itself has met regularly and has been heavily concerned in the stock assessments, at first in co-operation with the Committee of Three and with FAO, and since June 1969 independently. During this time there has been a notable increase in the part played in the work of the Committee by scientists specialising in population dynamics. The Committee still consists predominantly of scientists appointed by the Commissioners from their own national delegations, although it continues to have participation from FAO and more recently from individual scientists whom it invites to assist it in its work. Over the years the Committee has grown to a membership of over 30 scientists, and is finding it necessary to meet two or three times a year, for a total of about four weeks.

Organisation and finance: Before 1972 the Commission had only a part-time secretary and part-time office staff, provided by the UK Ministry of Agriculture, Fisheries and Food. In 1972 the Commission, in response to a recommendation from the UN Conference on the Human Environment, recognised the need to strengthen its organisation, and particularly its scientific capability. As a result, in 1976 the Commission established its own office in Cambridge, England, with a full-time scientifically qualified secretary and adequate supporting staff.

To finance its operations, the Commission has progressively increased its annual revenue from about £7000 in 1973-74 to about £130,000 in 1978-79. The bulk of the revenue (95 per cent) is derived from contributions from the member nations on a scale based on a flat rate (50 per cent) plus an 'areas of interest' component (30 per cent) plus a component for weight of whales caught (20 per cent).

Catch limits and protection of stocks: The Commission established some catch limits at the time of its establishment, but these originally were restricted to the Antarctic catches, and were applied, not to the individual species, but to the commercially important species of baleen whales as a group. This was done by setting quotas in terms of the blue whale unit. This unit consisted of 1 blue whale or 2 fin whales, or 2.5 humpback whales, or 6 sei whales. It was first introduced by the Bureau of International Whaling Statistics in 1939 as a means of expressing the catches of the various species in terms of oil yield. Its adoption by the International Whaling Commission in 1946 as a basis for regulations underlines the fact that stabilisation of output was at least a major consideration in early catch restrictions.

The Commission set its initial quota at a lower level than had prevailed during the 1930s. In 1946-47 it was 16,000 blue whale units, although catches before the war had been averaging about 24,000 units, and had reached nearly 30,000 units in 1938-39. The quota was held at 16,000 units until 1953 and thereafter fluctuated around 15,000 units until the mid-1960s. There were several years during this period in which the Commission failed to agree on a quota. In theory each country was then free to take any number of whales it wished, but in practice countries generally kept catches down to about the level of the previous season. At a special meeting in 1965 the Commission agreed that Antarctic catches of fin and sei whales should be progressively reduced so that by the 1967-68 season they would be below the sustainable yield of the 1967-68 stocks. The adoption of this principle was intended to ensure that no further reduction of the stocks took place. This decision immediately led to a major reduction in quotas, a process which has continued somewhat erratically until the present time.

As early as 1964 the Committee of Three had pointed out to the Commission the biological unsuitability of the blue whale unit, and the need for regulation species by species. This advice was reiterated frequently by the Scientific Committee. Nevertheless, species quotas were not introduced until 1969, when catch limits were set on fin and sei whales in the north Pacific for the first time. In the southern hemisphere the blue whale unit was not abandoned and species quotas introduced until 1972-73.

Specific quotas for sperm whales were first introduced in 1970 in the north Pacific, and in 1971 in the southern hemisphere. They were set separately for the two sexes in the following year, and in 1973 separation by areas began. Since that time separate quotas have also been set by areas for those baleen whales which could be taken in the southern hemisphere.

Parallel with this progressive application of smaller and more finely defined quotas has been the extension of protection to the more seriously depleted stocks. Right, bowhead and gray whales have been protected since 1931; blue and humpbacks have been protected from 1963 onwards in the southern hemisphere, and from about the same time in the north Atlantic. They have been protected in the north Pacific since 1966. Since 1975 a large number of other stocks have been protected as a result of new management practices then introduced.

New Management Procedure: In 1972 the United Nations Conference on the Human Environment passed a resolution calling for a 10-year moratorium on commercial whaling. A resolution which would have put this into effect by setting all quotas at zero was defeated at the annual meeting of the International Whaling Commission in that year. In the following year a resolution to end all commercial whaling within three years at the latest obtained a majority in the Commission, but not the three-quarters necessary to amend the Schedule.

In 1974, the same resolution was introduced, but was replaced by an amendment put forward by Australia. This amendment, which was carried by the Commission, laid down in principle a set of formalised rules to be applied by the Commission, on the basis of advice from the Scientific Committee, in determining which stocks should be protected, and what the catch limits should be for those whose exploitation was allowed.

Detailed rules, based on these principles, were presented to the Commission by the Scientific Committee in 1975, and were adopted. The effect of these rules, commonly referred to as the New Management Procedure, has been substantially to remove the Commission's decisions on catch levels from the domain of political negotiation. The Commission has, from 1975 onwards, almost invariably accepted specific recommendations on catch limits by the Scientific Committee. The only major exception has been with regard to the non-commercial fishery for bowhead whales in USA.

The new rules divide stocks into three categories:

Initial Management Stocks, which may be reduced in a controlled manner to the MSY level,[(1)] or some other optimum as this is determined;

Sustained Management Stocks, which are to be maintained at or near MSY level (or optimum, as this is determined);

Protection Stocks, which are below the Sustained Management level, and should be fully protected.

On Initial and Sustained Management Stocks whaling is allowed subject to the advice of the Scientific Committee, to an extent which should bring them to, or maintain them at, MSY or optimum level without risk of reducing them below it.

Initial Management Stocks are at present defined as those more than 20 per cent above MSY level, and Protection Stocks as those more than 10 per cent below this level. Quotas are set at 90 per cent of MSY for all stocks at MSY level or above, and graded linearly from this to zero at the boundary with the Protection Stocks.

The Commission has so far not been able to find a definition of optimum level better than the MSY level for this purpose. It has also decided for the present to use as MSY level that population level which gives the maximum number of animals in the sustainable catch rather than the alternative of using the level which gives the maximum sustainable weight.

The current procedure is throwing a very heavy load on the Scientific Committee, both because of the need to examine a large number of stocks in detail every year, and because of the requirement to make definite quantitative statements about stock and yield, when so often only very imprecise data are available. At the present time, for example, the Scientific Committee recognises about 50 stock units on which it must make some statement. Of these about 30 are in the Sustained or Initial Management categories, so that quotas have to be defined for them. Even for these

---

(1) The MSY (maximum sustainable yield) level is the population size which allows the greatest catch of whales each year without the population declining (so that whales taken by whalers or dying naturally are being balanced by births). These concepts are discussed in detail in Chapter 4.

stocks it is possible in only about two-thirds of the cases to make any quantitative statement as to the probable stock size and MSY. For the others, the Scientific Committee has to make a recommendation which it believes is in accordance with the guiding principles, but is based only on the apparent effect, if any, of previous exploitation. To assist in dealing with these situations important supplementary rules are applied.

Under the first of these, when a stock has remained stable for a considerable period under a regime of approximately constant catches, it is classified as a Sustained Management Stock, even though it is not possible to estimate the stock size itself. A catch quota is then set at about the average level of the previous catches.

Under the second, and perhaps even more important rule, exploitation of a currently unexploited stock is not allowed to begin until an estimate of stock size has been obtained which is satisfactory to the Scientific Committee.

A third rule provides that the annual catch shall not exceed 5 per cent of the initial exploitable stock size unless there is positive evidence that a larger catch can safely be taken.

Introduction of the New Management Procedure has had the effect of increasing substantially the number of stocks which are now protected from exploitation, and of reducing the catch quotas for most of the others. The fin and sei whales are now (for the 1978-79 season) protected throughout the southern hemisphere and north Pacific, and the fin whale is also protected in most of the north Atlantic. The rule that previously unexploited stocks may not be harvested until satisfactory stock assessments are available has also led to protection being applied to Bryde's whale in the southern hemisphere, and to minke whales in part of the north Pacific.

Under the Convention, the Commission is debarred from allocating quotas to particular countries or operations. It can only therefore fix quotas by geographical areas. Until recently the whaling nations were in direct competition for their shares of the quotas. Since the introduction of quotas by smaller geographical areas, however, the nations interested in operating in each area have negotiated the allocation of the quotas between themselves outside the Commission.

Size limits: When the International Whaling Commission was first established it immediately set minimum size limits for most of the species of whales which were of concern at that time. In the pelagic fisheries these were: for blue whales 70 ft; fin whales 57 ft (southern hemisphere), 55 ft (north Pacific); humpback whales 35 ft; and for sei and Bryde's whales 40 ft. It also set size limits for sperm whales, which recognised the difference between the sexes. These were, for males 38 ft, and for females 35 ft. In most cases the Commission also set rather lower size limits for coastal operations on the same species. There does not appear to have been any very specific reasoning behind the limiting sizes which were selected. In general they were fairly close to the sizes at sexual maturity for the various species, but they were also designed to allow the capture of most animals which were large enough to be economically worth taking.

No size limit has yet been set for minke whales in any areas.

In 1972 the development of more sophisticated models for sperm whales led to a reduction in the size limit for this species to 30 ft.

All these size limits refer to the minimum size of animal which may be taken. The extension in recent years of the sperm whale fishery into lower latitudes, where it operates on the breeding herds, has raised the question of whether disturbance of these herds, and particularly removal of the large harem masters, could be having an adverse effect on reproduction and therefore on recruitment. In the past two years therefore regulations have been introduced which prohibit the taking of males over 45 ft - the socially mature animals - in the areas between 40° South and 40° North during the breeding season in the respective hemispheres.

Closed seasons: Nearly all the fisheries under International Whaling Commission control are limited by a closed season. The Schedule either defines a particular period of operations, or requires each country to nominate a period of not more than a specified length as its season of operations. The practical effect of these limited seasons may be greatly modified by the operation of catch limits. In general the existence of a limit on the number of animals which may be taken removes much of the point in having a closed season.[1]

Australian policy and legislation

The Commonwealth Government's present policy on whaling is:
(a) Support for conservation of whales based on the advice of the Scientific Committee of the International Whaling Commission.
(b) Encouragement of all non-member countries to become members of the International Whaling Commission and strengthening the effectiveness of the International Whaling Commission's supervision of whale harvesting.
(c) Moratoriums on harvesting whale stocks which are endangered.
(d) Development of a more effective monitoring system of whale populations (Commonwealth Government, 1978, p.49).

The only State which has had a whaling operation in recent years takes a similar position:
The Western Australian Government believes that whales are a renewable resource which should be available for exploitation by man provided such exploitation is controlled. The Government has accepted the International Whaling Commission as the proper authority to determine the allowable levels of exploitation within the context of rational use of this renewable resource (Western Australian Government, 1978, p.4).
On the other hand at least the New South Wales and South Australian Governments have indicated that they would support a ban on commercial whaling in Australian waters (NSW National Parks and Wildlife Service, Inquiry Hearings, p.299; South Australian Government, 1978).

There is however agreement that the smaller cetacea, including dolphins and porpoises, should be generally protected by Australia, and this extends to their injury and mistreatment, with provision for licences for their capture for purposes of display or research.

(1) The Inquiry wishes to thank Dr K.R. Allen for the above account of the development of the International Whaling Commission.

It is appropriate then to examine the Commonwealth legislation so far as it embodies Australian policies upon whaling and also measures which are found in the legislation of both the Commonwealth and the States for the control of whales and whaling.

Reference has already been made to the 1931 Convention for the Regulation of Whaling. The Whaling Act 1935 was enacted to ensure the application of the provisions of that Convention. There were two main features of the Act, first it applied to baleen whales only and second, it was necessarily expressed to extend to Australian waters beyond territorial limits, in Australian territorial waters where proclaimed by the Governor-General, and also wide categories of ships (section 4.1). As Australia's first measure of protection an absolute prohibition was placed upon the taking or killing of any right whale, any calf or suckling whale or immature whale, or any female whale accompanied by a calf or suckling whale (section 6).

Similar statutes were enacted by the whaling States, Queensland and Western Australia, these being respectively the Whaling Act 1935 and the Whaling Act 1937. The Queensland Act has since been repealed. The Western Australian Act is still in operation. It applies within territorial waters, and in addition to any ship under Western Australian jurisdiction used for whaling. Cheynes Beach has been granted current licences under the Act both for its catchers and for the Albany station.

The next Commonwealth statute was the Whaling Act 1948 which made provision for the International Convention for the Regulation of Whaling 1946 to be brought into effect. Its main feature was to widen the application of the Act in accordance with the 1946 Convention by including in the definition of whale the sperm whale or any other prescribed species of whale.

In 1960 the Commonwealth enacted the Whaling Act 1960 for the regulation of whaling in Australian waters. This Act, which in section 4 repealed the earlier statutes, was designed to embody Australia's continuing commitment to the 1946 Convention, and also the principles which formed part of that Convention by virtue of the Protocol to the Convention of

1956 and amendments made from time to time to the Schedule to the Convention. The Whaling Act 1960 is included in Appendix 7.

The extended operation of the previous legislation to include all baleen whales, and also the sperm whale and any other whale of a prescribed kind is maintained in the 1960 Act (section 5.1). The Act is given a wide operation in the seas surrounding Australia and beyond the territorial limits of a territory that is not part of the Commonwealth, such as the Australian Antarctic Territory, but as before not in State territorial waters unless proclaimed (sections 5(1), 8(1)). Reference will be made later to the limits of the jurisdiction of the Commonwealth in the exercise of this power. The provision for the regulation of whaling to be found in Part II of the Act is more flexible than that adopted in the earlier legislation. Detailed provisions are not found in the Act itself. Instead the Minister is given the power to prohibit the taking or killing of whales by a notice published in the Gazette. The prohibition may apply at all times or during a period specified in the notice, and may cover all whales or whales of a species, kind or sex specified in the notice. It may also prohibit taking or killing by a specified method or equipment, or protect whales not exceeding a specified size (section 10(1)). Contravention of a notice is an offence punishable by fine (section 15).

Those powers, in relation to taking or killing of whales in waters other than Australian waters, may only be exercised to the extent necessary to give effect to the International Whaling Conventions[1] (section 10(5)). Control of whaling by means of licensing, as instituted by the earlier legislation, is maintained; forms of licence are included in Appendix 7. In the present Act, the power is given to the Secretary to the Department of Primary Industry (section 11(1) and (2)). Again a flexible approach, which was no doubt designed to cater for changes from time to time in the regulations adopted by the International Whaling Commission, appears in the omission of any specific conditions to be included in the licence. This matter is left to the Secretary, who may specify conditions in the

---

(1) Under the Act 'the International Whaling Conventions' is defined to mean the 1946 Convention, the Protocol, and any amendment to the Schedule to the Convention under Article V, being an amendment which has become effective with respect to Australia (section 5(1)).

licence (section 12). But the Act does provide for the range and nature of the conditions to be imposed. Such conditions are to be included as the Secretary considers necessary to give effect to the International Whaling Conventions, and conditions relating to the taking or killing of whales in waters other than Australian waters are not to be specified in a licence except for the purpose of giving effect to the International Whaling Conventions (section 12(2) and (3)). Provision is made for research and development in Part IV but, as will appear, it is limited in operation.

It can be said that the Act does give adequate powers for the control of whaling within Australian waters. No submission has been made to the contrary. The Department of Primary Industry has had no problems arising from the scope of the Act. However, in relation to the powers conferred in respect of research under Part IV - Research and Development, it seems that the research needs to be related to whaling on a commercial basis or its development in Australian waters. Accordingly, if it becomes the basis of Australia's policy to put an end to whaling, it may be desirable to widen the power to make it clear that it is not restricted to research for commercial purposes. However other powers exist for the conduct of scientific research, and specifically, under the National Parks and Wildlife Conservation Act 1975, which will be referred to later, there is ample power under section 16(1)(f) to conduct research relating to the protection, conservation and management of wildlife (which is defined to include cetacea).

Without referring to the whaling notices in detail, it is sufficient to say that the powers under the Act have been used to provide for matters such as the minimum size limits for the taking of whales, closed seasons to cover the breeding period and a total prohibition on the taking of female whales accompanied by calves or suckling whales. The taking of the right whale and, since 1964 and 1968 respectively, of the humpback whale and the blue whale has also been prohibited (see Commonwealth Government, 1978, Appendix A).

In 1975, an amendment to the Whaling Regulations (see Appendix 7) was made to include, for the purposes of the definition of 'whale' as contained in the Act, further species of Odontoceti, in effect, the pilot whale, the pygmy sperm whale and all species of beaked whales, dolphins and porpoises found in the southern hemisphere. In fact dolphins and porpoises have never been the object of a direct commercial fishery in Australia. This amendment of the regulations may have been directed to controlling the taking of small whales for use in oceanariums bearing in mind the growing public interest, both in Australia and other parts of the world, in such displays. To date no licences have been issued under the Whaling Act for the taking of these smaller species. Permits may be granted for the taking or killing of whales for scientific purposes (section 26), and this power is broad enough to cover studies of animal behaviour.

Public concern has been expressed, particularly by conservation groups, about the need for a wide range of protection measures for the small species (see Knott, 1978). Certainly there are no comprehensive provisions to be found in the Act which relate generally to the killing, capture, injury or mistreatment of such species or their treatment during transport and while kept for live display purposes. If the Act were required to deal with these problems, substantial amendment would be required.

The only other Commonwealth legislation applicable to cetacean conservation is the National Parks and Wildlife Conservation Act 1975 which establishes the Australian National Parks and Wildlife Service (ANPWS). The functions of the Service's director include protection, conservation, management and control of wildlife. The director is also empowered to conduct surveys and collect statistics about animals and plants and to carry out research and investigations relevant to the protection, conservation and management of wildlife (section 16(1)). Because of the wide definition of 'wildlife' contained in the Act, it is clear that all species of cetacea which are indigenous to the Australian coastal sea, the continental shelf of Australia or the superjacent waters or which, being migratory animals, visit those waters periodically or occasionally, are included in the definition of 'wildlife' (section 3(1)).

The Service's administrative responsibilities extend to certain international conventions including the Convention on International Trade in Endangered Species of Wild Fauna and Flora, which came into force for Australia on 27 October 1976. The ANPWS is the scientific authority and the Bureau of Customs is the management authority for the operation of the Convention in Australia. The Customs (Endangered Species) Regulations regulate the import and export of specimens and products of endangered species listed in the Appendixes to the Convention. Some cetacea are covered by the Convention.

Since its establishment in 1975, the Service has commissioned several research projects and surveys related to dolphins and large whales, dugong and turtles. A consultancy report on development of policy for marine reserves in Australia has also been completed recently.

The scope of the National Parks and Wildlife Conservation Act is extended by the broad powers for the making of regulations under section 71 of the Act. Regulations have been made covering offences relating to wildlife in parks or reserves proclaimed under the National Parks and Wildlife Conservation Act 1975.

In all States of Australia and the Northern Territory, statutes relating to the conservation of wildlife have been enacted. Draft legislation is being prepared for the Australian Capital Territory. The other relevant legislation is:
- National Parks and Wildlife Act 1974 (New South Wales);
- Wildlife Act 1975 (Victoria);
- National Parks and Wildlife Act 1975; Fauna Conservation Act 1974 (Queensland);
- National Parks and Wildlife Act 1972 (South Australia);
- Wildlife Conservation Act 1950-75 (Western Australia);
- National Parks and Wildlife Act 1970 (Tasmania);
- Territory Parks and Wildlife Conservation Ordinance 1976 (No.23 1977 as amended) (Northern Territory).

It should be noted that 'wildlife' as defined in the Tasmanian National Parks and Wildlife Act 1970 (section 31(c)) excludes animals defined as 'fish' under the Tasmanian Fisheries Act 1959 as amended. Thus whales are dealt with under the latter Act. Under the Queensland Fisheries Act 1976, marine mammals (that is including whales) are included in the definition of 'fish' and are included in the list of protected species under that Act (Second Schedule).

In the statutes of New South Wales, Victoria, South Australia and Western Australia, and the ordinance of the Northern Territory, the term 'wildlife' has been defined in terms broad enough to include all cetacea. However, the Wildlife Conservation Act 1950-75 of Western Australia excepts from its general prohibition on the taking of wildlife the taking of whales under licence pursuant to the Whaling Act 1937 of that State.

Following High Court decisions on the Seas and Submerged Lands Act affirming Commonwealth sovereignty seawards from the low water mark, the Commonwealth and State Governments are discussing responsibilities for matters relating directly to marine parks and reserves and marine wildlife. It is understood that the aim is to resolve jurisdictional problems in this area by mid-1979.

The Constitutional position is that State power with regard to fisheries, which would include cetacea, extends at least to the three-mile limit of the territorial sea (Pearce v. Florenca (1976) 135 CLR 507). Generally speaking, the existing legislative arrangements between the States and the Commonwealth in relation to fisheries are that State legislation applies to fishing out to three miles and that Commonwealth legislation applies beyond that limit. Following a Premiers' Conference held on 22 June 1978 the Attorney-General announced that joint Commonwealth-State authorities would be established for off-shore fishing. These authorities would, by agreement of the Commonwealth and the States, manage specified fisheries from the low-water mark out. By agreement some fisheries outside the territorial sea could be allocated to the States concerned. Apart from Western Australia, the States have not been concerned with whaling.

For effective legal control to be achieved over the capture and treatment generally of cetacea, and particularly dolphins, similar discussions will be necessary. It would seem that whatever arrangements are reached for territorial waters, complete protection for cetacea in all Australian waters can be achieved only by the enactment of further legislation by the Commonwealth. If there is a change in whaling policy, consideration should be given to the repeal of the Whaling Act 1960 and its replacement by legislation similar to the marine mammals protection Acts of the United States of America and New Zealand.

Under legislation of this nature the taking of marine mammals and the import or export of marine mammals or their products is prohibited without a permit from the designated authority. In the United States the Act provided for a moratorium on commercial whaling. By general executive direction permits are limited to purposes such as scientific research, public display, aboriginal subsistence, or accidental killings incidental to commercial fishing operations (US Public Law 92-522; Connecticut Cetacean Society, 1978). The New Zealand legislation also provides for permits for similar purposes.

Administration of whaling policy

Inspection and infractions: As indicated earlier, the International Whaling Commission requires that all whaling operations be inspected by the government concerned and all the infractions of regulations reported to the Commission. The Commission also must be supplied with specified information on catches and catching operations and selected samples from each whale must be available for scientific investigation.

The Commonwealth Department of Primary Industry and the Western Australian Department of Fisheries and Wildlife have a working arrangement covering the inspection and reporting of whaling activities at the Albany station. Officers of the Western Australian Department are appointed as whaling inspectors under the Whaling Act 1960. Information is recorded on the type of whale, its measurements and sex. If a female, the mammary glands are examined. Inspectors also ensure the collection of specimens for research purposes and forward them to the Western Australian Museum. Catch rates and details of any infractions, including gunners' reports, are forwarded to the Department of Primary Industry weekly. Officers of the Western Australian Department have also been appointed for certain periods as international observers at whaling stations in South Africa and Brazil.

The infraction rate by Cheynes Beach was below 0.5 per cent for the years 1973 to 1976. However in 1977 this rate increased to 4.5 per cent when 28 whales were taken which were longer than the newly introduced size limit of 45 feet allowed during the breeding period from October to January. Notwithstanding the report by gunners that they found difficulties judging when the whales met this size restriction, the Commonwealth Government reduced the Cheynes Beach quota for 1978 by 28 whales, thus penalising the Company while also affording added protection to the stock. In accordance with the International Whaling Commission's Schedule no bonus was paid to the gunners or crews of the whale catchers in respect of the taking of these whales.

Departmental responsibility: Australia has been an active member of the International Whaling Commission, particularly in recent times when Australia has chaired the Commission and Australian scientists have occupied the positions of Chairman and Vice-Chairman of the Scientific Committee. The Australian Commissioner, who has generally been the head of the Fisheries Division of the Department of Primary Industry, has usually been accompanied by Australian scientists and also industry representatives. In 1970 and 1977 the director of the Western Australian Department of Fisheries and Wildlife was also included in the delegation.

In recent years increasing national concern for the conservation of whales has been reflected in the composition of the Australian delegation. An officer of the Department of Environment joined the delegation in 1973 when the question of a moratorium first arose on the agenda. More recently, officers of the Australian National Parks and Wildlife Service were on the Australian delegation to the 1977 Commission meeting, and the director of the Service attended the special Commission meeting held in Tokyo in December 1977 and the 30th Annual Meeting of the Commission in London in June 1978. Representatives of Project Jonah were included in the delegation as observers at the Tokyo and London meetings.

If Australian policy on whaling is to be changed to become more directed towards protection, the question will arise whether administration of this policy should be the concern of the Minister responsible for the environment. It has been suggested that the Australian National Parks and Wildlife Service, with its role as an adviser to the Government on nature conservation and wildlife policies, would be the appropriate body to undertake this task. It may also be argued that the Department of Primary Industry, having built up considerable expertise on whales and whaling and having established excellent relations with scientists active in the field and the various national delegations to the Commission, would be in the better position to carry the work forward. We make no recommendation on this issue, which is a matter of government administration.

Australia's powers in relation to whaling within the 200-nautical-mile fishing zone

During 1978 the Government introduced the Whaling Amendment Bill 1978 to amend the Whaling Act 1960. As Mr Sinclair, the Minister for Primary Industry said, in moving the second reading of the Bill, it contains significant amendments providing for 'the legal machinery to implement the latest management measures approved by the International Whaling Commission for the taking of whales. The measures are provisions for the prohibition of the taking of whales of a specified species, kind or sex in excess of a certain size or number.' But the main purpose of the Bill was to extend Australian control over whaling to cover the 200-mile fishing zone as defined by the Fisheries Act 1952-78 section 4. The Minister said, 'This Bill effects an extension of the Government's jurisdiction with respect to whaling by much the same means as the Fisheries Amendment Bill (now the Act), the difference being that, whereas the Fisheries Amendment (Act) provides for exclusive control over foreign fishing vessels within the 200-mile zone, the Whaling Bill confers a jurisdiction which is in some respects qualified. The qualification arises from the Conventions and the effect is to remove from Australian whaling jurisdiction, vessels or aircraft which are flying the flag of, or registered in, a foreign country that is a party to the International Whaling Conventions and whose use in connection with whaling is duly authorised by that country and is not in contravention of any of the provisions of the Schedule to the Convention'. The provisions of the Bill are certainly consistent with Australia's present policy in support of whaling and the conservation of whales based on the advice of the Scientific Committee of the International Whaling Commission. It is to be noted that if passed into law the amendment would permit a foreign whaler to take whales within the 200-mile Australian fishing zone without an Australian licence, provided the whaler did so under the authority of its flag state which was a member of the International Whaling Commission and also in conformity with the provisions of the Conventions. The relevant provisions of the Whaling Amendment Bill are sections 5(e), 7(b), (c), (d).

Within Australia the Amendment Bill has been the subject of controversy and has been strongly opposed by the various conservation groups which take the view that the cessation of whaling in Australia which they strongly support can be made effective only by legislation prohibiting foreign whaling within the 200-mile fishing zone. Their argument is that it would be an intolerable situation, now that Cheynes Beach has decided to close down its operations, if Australia decided to adopt an anti-whaling policy, for its quota in Division 5 to be taken up by other countries and availed of by whaling within the 200-mile zone. In the meantime the passage of the Bill has been delayed pending the completion of the Inquiry and the consideration by the Government of this report.

There are two questions of a legal nature which arise for consideration here. They are, first, whether Australia has the power to control and in particular to prohibit foreign whaling within the proposed 200-mile fishing zone, and second, whether the exercise of that power is consistent with Australia's obligations as a party to the Conventions.

The first of these questions raises issues both of Australian constitutional law and of international law. Upon these matters an opinion has been obtained from the Attorney-General's Department. The department advised that the principal sources of Commonwealth power to regulate whaling were the 'trade and commerce' power (Constitution, section 51(i)), the 'fisheries' power (section 51(x)), the 'external affairs' power (section 51(xxix)) and the 'incidental' power (section 51(xxxix)). With respect to the 'fisheries' power, the view was expressed that the term 'fisheries' comprehended whaling and also that the term included both the taking and conservation of fish. From this we conclude that section 51(x) empowers the Commonwealth to legislate concerning the taking and conservation of whales. The 'fisheries' power extends to 'Australian waters beyond territorial limits'. The department advised that the term 'Australian waters' was considered by the High Court in the case of Bonser v. La Macchia (1968-69) 122 CLR 177, and was recognised to have a very extensive meaning, not necessarily confined to the limits already proclaimed under the Fisheries Act which in some instances extended more than 200 miles from the coast.

Turning to the position at international law, the advice given was that international law now recognises that coastal states have the right to establish 200-mile fishing zones off the coast, subject to certain qualifications which for all practical purposes are conveniently set forth in the Informal Composite Negotiating Text (ICNT) that has emerged from the current United Nations Law of the Sea Conference. The ICNT, though still a negotiating text, reflects in its provisions with respect to fishing zones the current state of customary international law on the subject. In article 65 of the ICNT the international community recognises the right of coastal states to prohibit, regulate and limit the exploitation of marine mammals. That article also provides that states shall co-operate either directly or through appropriate international organisations to protect and manage marine mammals. In the opinion of the Attorney-General's Department, the power expressed in article 65 would enable a coastal state to control and prohibit foreign whaling within the 200-mile fishing zone. The Whaling Amendment Bill would enable Australia to exercise this power. As noted, however, the power would not be exercisable, if the Bill becomes law, in respect of the ships or aircraft of countries, other than Australia, that are parties to the International Whaling Conventions.

In a later opinion the Attorney-General's Department has advised that, upon a consideration of the provisions of the International Whaling Conventions there is nothing in these Conventions that would prevent Australia from prohibiting foreign whaling within its 200-mile fishing zone.

These questions were adverted to by Project Jonah, and it obtained for the assistance of the Inquiry a legal opinion from Professor D.H.N. Johnson, Professor of International Law in the University of Sydney, New South Wales. Professor Johnson, in his opinion, concluded that a coastal state, whether or not it is a member of the International Whaling Commission, is entitled to prohibit whaling by foreigners in its 200-mile zone. Dr K.D. Souter, senior tutor of St John's College, University of Sydney, whose specialty is international law and international relations, took the same view (Inquiry Hearings, pp.205-10).

The Inquiry agrees with the conclusions reached in these opinions. That the coastal state has the power to prohibit whaling within the 200-mile fishing zone is clearly the opinion of the United States, Canada and New Zealand, each of which has taken action on this matter.[1] Accordingly, if this Inquiry takes the view that Australia should adopt a policy against the continuation of whaling it would be consistent with that view for the Inquiry to recommend that whaling by foreign ships in the 200-mile fishing zone should be prohibited.

With respect to the Australian Antarctic Territory, no statement has been made by the Australian Government as to whether or not a 200-mile fishing zone will be proclaimed. However, along with the other Antarctic Treaty consultative parties, Australia is engaged in the negotiation of a Convention on the conservation of Antarctic marine living resources. The Convention in its present draft form applies to the Antarctic marine living resources of the area south of 60° latitude and also to the Antarctic marine living resources of the area between that latitude and the Antarctic convergence which forms part of the Antarctic marine ecosystem. It is intended therefore that the Convention will cover whales, but not in any way which would be contrary to arrangements under the International Whaling Convention. The draft Convention contains a provision that nothing therein contained shall derogate from the rights and obligations of contracting parties under the International Whaling Convention.

---

(1) For the United States, see the Marine Mammal Protection Act of 1972 (as amended), section 102(f), and the Fishery Conservation and Management Act of 1976, sections 3(8) and 101; for New Zealand, see the Marine Mammals Protection Act 1978. Canada's whaling regulations prohibit, with certain exceptions, whaling in Canadian fisheries waters which extend to 200-miles (Mercer, 1978).

Australia's future whaling policies will therefore need to be taken into account in its Antarctic Treaty consultations and also in relation to the 200-mile fishing zone should Australia decide to declare such a zone in relation to the Australian Antarctic Territory.

4. THE BASIS OF SCIENTIFIC ADVICE

Whales migrate far beyond national boundaries and therefore effective management - which may range from full protection to controlled whaling - requires international co-operation. This is complicated by the seasonal concentration of many whales in the Antarctic where national and international rights and responsibilities are still being defined.

Some nations regard whales as a renewable resource which they are free to exploit; others are opposed to the killing of whales. So an internationally accepted policy, prudent enough to meet the major concerns of nations opposed to whaling, is needed for the regulation of whaling. It is against this background that we review the basis of the scientific advice provided to the International Whaling Commission. This chapter provides a description of the general theory behind the present management procedure, and the way in which this is used to fix catch limits, and sets out out the major criticisms which have been made of both the theory and its application.

The basic question is how many whales may be taken 'safely' from a particular stock. Simple theories of population growth suggest that at any given population level a certain number of whales can be taken each year without reducing the population - whales which are taken by whalers or die naturally being balanced by births. This however is looking at the stock in isolation. In particular it assumes that exploitation of the stock does not lead to other changes in its environment (such as the expansion of other species using the same food source) which could reduce its drive to return to its initial level of population.

Management of whaling under the International Whaling Commission's New Management Procedure is based substantially on the supposition that the general theory outlined above can be put into practice, so that a stock can be brought to the level at which the annual yield is largest, and with some caution this 'maximum sustainable yield' (or at least a major part of it) can then be taken every year without further reducing the stock.

As seen in Chapter 3, the New Management Procedure requires estimates of both the current stock size and the maximum sustainable yield level.[1] It is not a simple matter to obtain these estimates; direct counts are not feasible - except for a few stocks such as the bowhead where some confirmatory counts can be made as the whales pass close to land - and only a limited amount of information can be obtained from recoveries of whale marks fired into the body of the whale. For most stocks the only practicable approach is an indirect one based on mathematical models which try to reflect the biological behaviour of the stock. Most of these models depend substantially on the use of whaling statistics.

Assessments of the state of individual whale stocks can also be based on data from sightings such as those provided by scouting boats or aircraft associated with a whaling fleet. This is particularly useful for species or stocks which are protected, where catch data do not exist and there is naturally no recovery of internal whale marks. For other stocks sightings are mainly used as a check on the analyses based on catch data.[2]

---

(1) The maximum sustainable yield level is usually taken on theoretical grounds to be a certain percentage of the initial stock size (before exploitation), and it is the initial stock size which is estimated.

(2) More extensive discussions of the available techniques for assessing the state of a whale stock can be found in Gambell (1976); Allen and Chapman (1977); Holt (Inquiry Report, Volume 2); and FAO/ACMRR (1978) The last two papers also provide a broader consideration of the current approach to whale harvesting.

The background to the FAO/ACMRR reference is of interest. This paper is the draft proceedings of the Scientific Consultation on the Conservation and Management of Marine Mammals and their Environment held in Bergen, Norway, in September 1976 under the auspices of the Advisory Committee on Marine Resources Research of the Food and Agriculture Organisation of the United Nations. The consultation was attended by nearly 200 marine mammal specialists, including many who were members of the International Whaling Commission Scientific Committee. It provided an opportunity for an extremely broad review, independent of the International Whaling Commission, of the status of whale stocks, their management and the research directed towards them. Appendix 1 of Holt (Inquiry Report, Volume 2) provides a series of extracts from the proceedings related to sperm whales.

Even from these introductory remarks it is clear that there are some fundamental difficulties associated with the advice developed by the Scientific Committee. These are among the matters mentioned by Dr Lee Talbot, Senior Scientist in the United States Council of Environment Quality, in discussions with the Inquiry. He said 'the generalised concept (of maximum sustainable yield) is elegant both in its simplicity and its adherence to the basic theories of the behaviour of the population. The problem is that as applied, it is totally unsound' (Talbot, 1978). It is thus important that the Inquiry examine carefully the assumptions made by scientists in order to reach conclusions or the degree of uncertainty and risk in the present management procedures.

The Scientific Committee itself recognises many deficiencies in the New Management Procedure and has established a working group to develop a better procedure. One approach suggested is to examine ways of setting catch limits on the basis of characteristics of the present stock rather than relying substantially, as now, on theoretical relationships between the present and initial stocks. This could be done for example by linking catch limits to the replacement yield, that is, the yield which can be taken without reducing the stock size. These ideas seem promising, but they need further development.

Maximum sustainable yield[1]

A population or stock of whales is generally assumed to have a natural maximum level within its environment, when additions by births are balanced by losses by deaths. If its numbers are reduced compensating factors come into effect, increasing the rate of reproduction in relation to mortality, and so tending to re-establish the population at its previous level; the opposite effect occurs if the population exceeds the capacity of its environment to sustain it. If a population could not respond in this way to changes in its density, continued reductions would drive it to extinction.

---

(1) This and the following section substantially follow Holt (Inquiry Hearings, pp.402-10).

For whales there is some evidence that pregnancy rates increase when population density is reduced, and there is circumstantial evidence that this is related to increases in the availability of food when the density of population decreases. While density is also expected to affect mortality rates, these rates are difficult to measure and it has not been possible to establish firmly that there is such a relationship.

When a population responds to a decrease in density there are more births than deaths from natural causes. If this net addition to the population equals the number taken by the whalers each year the population will be maintained at the reduced level and the number taken represents a sustainable yield.

The first ideas for sustainable yields from wildlife resources came from the consideration of the population growth curve which was formulated at the turn of the century as an 'S' shaped curve. A small population of animals growing within its environment will increase in numbers at an increasing rate at first, then at a reduced rate, and gradually it will reach a maximum determined by the carrying capacity of the environment (see Figure 4.1).

It was observed that as far as fish[1] were concerned, a crop could be taken at a particular time before the maximum level was reached; continually taking this small crop would hold the population steady. The yield that could be taken over a given time is shown in Figure 4.1 for one particular population size.

Only a very small yield can be continually taken from a population which is very near to its maximum level. Similarly, a very small population can only produce a very small sustainable yield. In the first case the yield is small because the growth rate of the population is small,

---

(1) Holt has drawn attention to a fallacious impression which had arisen in the literature that the concept of MSY arose in relation to fish populations and was not applicable to populations of whales. While the concept was taken up strongly for the purposes of fisheries, the idea started with whales, 'the first description of the so-called sigmoid S-shaped curve of population growth being interpreted as a sustainable yield model by a Norwegian. Its application was to the blue whale' (Inquiry Hearings, p.383A).

Figure 4.1

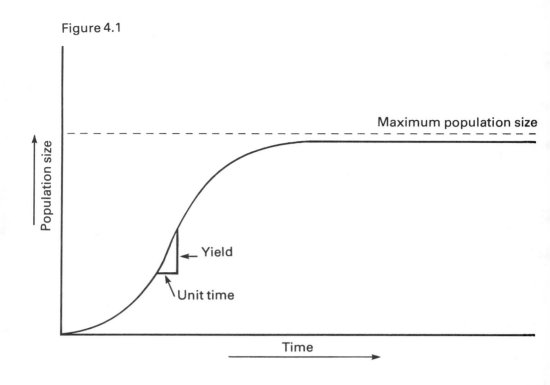

even though the population itself is large. In the latter case the yield is small because the population is small even though the growth rate is high.

Somewhere between these extremes there is a point where the largest possible yield is available because there is a fair response to the reduced density but still a fairly large population. This is the maximum sustainable yield (MSY) that could be obtained from the population. Figure 4.2 illustrates this.

The question then arises whether in practice population size can safely be held down by exploitation and to what extent. For simplicity of exposition it is convenient to assume that there is no change in the rate of deaths as population size is reduced and that the response to changes in density occurs only through changes in the pregnancy rate.

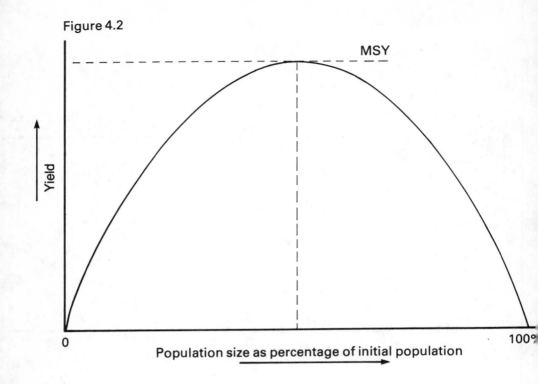

Figure 4.2

Figure 4.3 illustrates three hypothetical relationships between pregnancy rates and population size. If the reproduction rate increases to the point 'a' when the population is reduced, there would be substantially more births than deaths. If instead the pregnancy rate moved to a point 'b' there would be still more births than deaths and the sustainable yield would be higher. Alternatively, it could go to point 'c' and there would be a smaller difference and yield.

While the sustainable yield at a point such as 'a' may be fairly clear, it is more difficult to know what the population would do under more intensive exploitation. This would put it under much greater stress than it would normally face from natural changes in its environment, and it is difficult to know how the pregnancy rate would respond to greatly reduced population levels - there is the possibility that it may drop sharply.

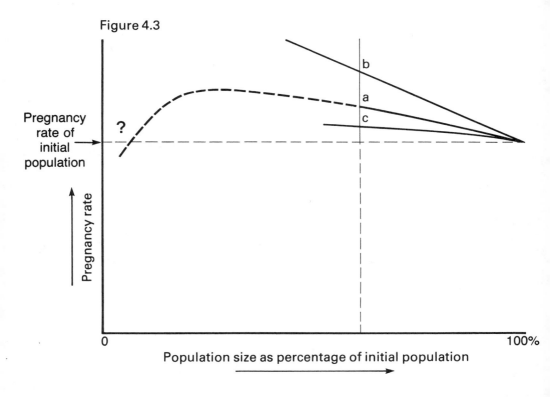

Figure 4.3

Application of maximum sustainable yield to baleen whales

As mentioned above, there is no direct evidence of changes in the natural mortality rate for whales corresponding to changes in the size of the population. It has generally been assumed in models of whale populations that the natural mortality rate, even for juveniles, is constant, and that the population response to changes in its density occurs through changes in the reproductive rate.[1]

The shape of the pregnancy rate response curve is important as it dictates where the maximum of the yield curve of Figure 4.2 occurs and hence at what population size the maximum sustainable yield is produced. The

---

(1) There are some exceptions. For example the sei whale model used at the 1977 Scientific Committee meeting incorporated a density dependent effect on juvenile mortality.

simplest models used for fish[1] and whales have a straight line pregnancy rate response and the curve of sustainable yield that this gives is the familiar one (Figure 4.4(a)). It is a symmetric curve whose maximum is at the point where the population is at half its natural value. At that point the difference between the reproduction and the natural mortality rates is about the middle of its range, the population is about the middle of its range, but the two multiplied together provide the maximum.

However, if the pregnancy response curve is not straight but bulges upwards, the corresponding sustainable yield curve has a maximum at higher than 50 per cent of the initial population size (Figure 4.4(b)). If the curve instead sags downwards, the corresponding sustainable yield curve has a maximum at less than 50 per cent (Figure 4.4(c)). The position of the MSY level is thus very sensitive to the shape of the pregnancy response curve.

There appears to be some limited evidence in general, and strong evidence in the case of fin whales, that the pregnancy response rule in whales is that illustrated in Figure 4.4(b).[2] This has led the Scientific Committee of the International Whaling Commission to adopt, somewhat arbitrarily, the level of 60 per cent as the maximum sustainable yield level for baleen whales, and a particular mathematical form for the pregnancy response relationship.[3] It is almost impossible in practice to determine the actual maximum sustainable yield level for a population, even if this

---

(1) In most models used for fishery management the natural mortality for juveniles is density dependent but not the natural mortality throughout the remainder of the animal's life. While the reproductive rate is also density dependent the major response in fish stocks studied is through these changes in juvenile mortality.

(2) Holt, Inquiry Hearings, p.408; see also Holt, Inquiry Report, Volume 2, pp.32-6.

(3) More details on the mathematical form for the pregnancy response relationship can be found in Appendix 8 on the sperm whale model.

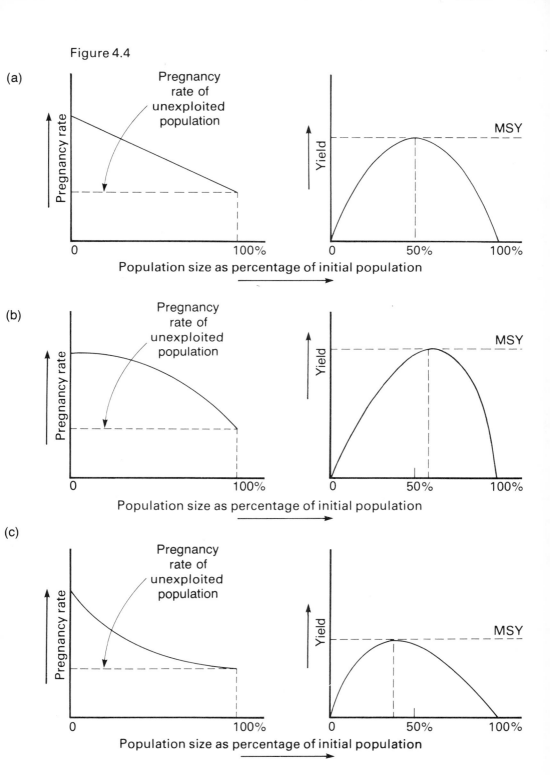

Figure 4.4

is approached indirectly by attempting to identify the changes in the factors which influence it (Holt, Inquiry Hearings, p.408).[1]

The sustainable yield is also closely linked to the size of the pregnancy response: the greater the response assumed the greater the corresponding sustainable yield (see Figure 4.5).

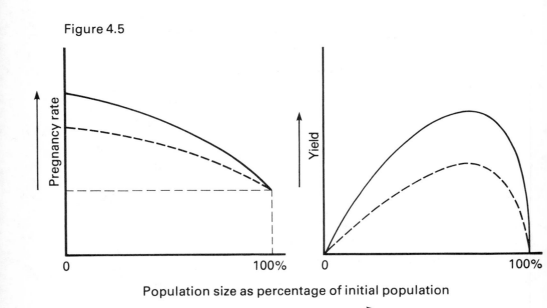

Figure 4.5

Population size as percentage of initial population

The discussion up to now has been somewhat simplified. In practice attention has to be focussed on the exploitable population - that is the age groups that whalers are interested in catching. The exploitable

---

(1) The modelling approach for sei whales developed at the 1977 meeting of the Scientific Committee incorporated a provision for an effect of changes in the density of blue and fin whales. This was prompted by indications of earlier maturity of sei whales and an increase in their pregnancy rates, possibly associated with an increase in resources available to them as a result of the decline in the numbers of larger whales.

population will take some time to respond to changes in its size, as
obviously it will be some years before the animals born as a result of an
increase in the pregnancy rate become part of the exploitable population
and can be counted as an increment from which a yield can be taken. This
introduces the need for a quite complex consideration of changes over
time in the composition of the population by age and sex. If the total
population has not reached its stable age and sex composition, the
number of young whales entering the exploitable population may be fewer
than anticipated, and unless this is taken into account in setting catch
limits, it may have serious long-term effects. This qualification applies
particularly to sperm whales because of their complex social structure.

Application to sperm whales

Because of the complexity of the model used for sperm whales, only a brief
account is provided in this chapter. A more extensive exposition is given
in Appendix 8.

As indicated in the discussion of reproduction and social order
among sperm whales in Chapter 1, sperm whales are polygynous. Each breeding
male has a harem during the breeding season. Males do not participate
in breeding until they are socially mature - by which time they have been
both exploitable and physically mature for some years - while females
mature much earlier, before even reaching exploitable size.

The population components used in the model can thus be listed:
(a) juveniles of each sex;
(b) sexually immature young of each sex;
(c) unexploitable mature females;
(d) exploitable mature females;
(e) exploitable socially immature males;
(f) socially mature males, with two elements: the number required
for breeding (in harems or as reserves), and the remainder
which are surplus to breeding requirements and hence exploitable.
The number of animals in each component each year is adjusted by allowing
for births, deaths (both natural and due to whaling) and transitions from
one component to the next.

Sperm whales are generally assumed to respond to changes in population density only through changes in the pregnancy rate following upon changes in the density of the mature female population. The pregnancy rate response curve is chosen in the same way as for baleen whales so that, if females alone were considered, the maximum sustainable yield level would be at 60 per cent of the initial size of the mature female population.

The maximum sustainable yield from a stock cannot however be estimated as directly for sperm whales as for baleen whales. Features such as the different ages at which males and females become exploitable and the polygynous habits of the species must be taken into account. The available yield must therefore be determined for males and females simultaneously to produce a maximum combined yield, with separate catch limits being set for each sex. Although the results naturally depend on the values chosen for the parameters, for most southern hemisphere sperm whale stocks the present estimate is that the total number of whales which can be taken annually is highest when the exploitable male population is reduced to 32 per cent and the exploitable female population to 76 per cent of their respective initial levels.

Use of maximum sustainable yield in the New Management Procedure

When catches are managed, a decision has to be made about the population size at which stocks should be maintained. In adopting the New Management Procedure (see Chapter 3) the International Whaling Commission has decided in principle to stabilise the population size of each whale stock 'near to' its maximum sustainable yield level (MSYL) pending any broader determination of optimum stock levels. In practice, as a conservative measure it sets catch limits that, if estimated accurately, would stabilise stocks at levels above MSYL. However, it does not apply the cut off at precisely MSYL, but currently defines 'near to' as between 20 per cent above and 10 per cent below MSYL and classifies this as the 'Sustained Management Stock' (SMS) range. Stocks at levels within this range could be held at their levels by taking their annual sustainable yield. Figure 4.6, which incorporates the 60 per cent MSYL curve of Figure 4.4(c), shows the sustainable yield curve generally adopted for baleen whale stocks,

Figure 4.6

Comparison of permitted catch limit under New Management Procedure with sustainable yield curve for population where MSYL occurs at 60% of initial level

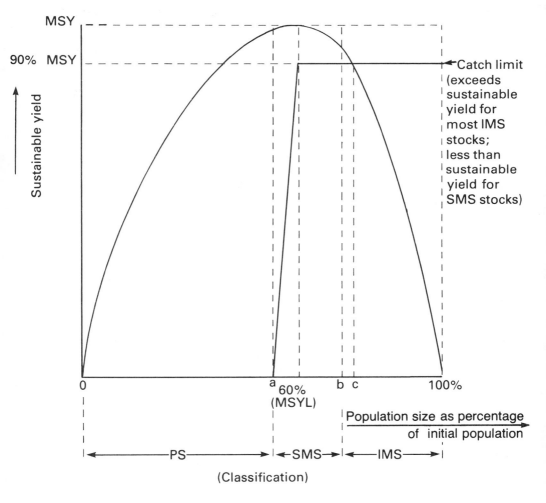

a (54%)   10% below MSYL } range of SMS category
b (72%)   20% above MSYL }
c (75.4%) point where 90% MSY catch limit equivalent to sustainable yield; population should stabilise at this level

together with the catch limits actually permitted by the New Management Procedure at different population levels.

It will be seen that over all the SMS range, the catch limits actually set are below the sustainable yields that have been calculated. The maximum catch limit of 90 per cent of MSY applies to any stock which is above MSYL. With this restraint, a stock above MSYL should stabilise at 75.4 per cent of the initial population size. For a stock in the SMS category but below MSYL, the permissible catch limit is well below the sustainable yield, especially if the stock is assessed as close to the Protection Stock category.

In the case of sperm whales a similar approach is adopted separately for males and females when classifying the stock and setting catch limits. The appropriate male and female MSY levels are naturally used as reference points rather than the 60 per cent reference point generally used for baleen whales.

Using the models

Not all whale stocks are modelled in the same way as there are special circumstances in some cases. While the following description of the use of the models is valid generally, it applies more directly to the sperm whale model. Further details of this particular model and its use can be found in Appendix 8.

Values must first be set for all the parameters of a model. Biological studies on whales taken provide most of the information used in choosing values for such parameters as the age at maturity, the natural mortality rate and the possible range for the pregnancy rate. Other available information, again generally collected in the course of whaling, is used to provide estimates for parameters such as the average number of females in a harem in the case of the sperm whale model.

The model needs a starting point from which to follow through the changes from year to year in each population component. It is assumed that at that starting point (usually 1946 for sperm whales) the population was both at the maximum carrying capacity of its environment and stable (so that movements in and out of each population component balanced each other). The number of whales in this initial population is then chosen, broadly speaking, so that the history of the population given by the model (making allowance for actual catches) reflects as closely as possible the changes actually observed in the abundance of whales.

This last point is best illustrated by a simple example. Suppose indications are that the number of whales is halved over a certain period. If the catch over this period was 10,000 whales, the population at the beginning of the period must have been 20,000 (ignoring for simplicity the effect of births and natural deaths).

Indications of changes in relative population sizes are usually based on catch per unit effort (CPUE) data - that is, in general terms, data indicating the success of whaling in relation to the amount of time ('effort') used. These data, which are usually recorded as 'catches per catcher day worked', are adjusted to allow for changes in the efficiency of whaling over the years such as the introduction of underwater scanning equipment (ASDIC), and, where possible, are used to obtain estimated figures for 'catch per searching hour' for each year over the relevant period. It is the series of CPUE data adjusted in this way which is used as the best indicator of changes in the abundance of whales over time.

One of the exceptions to this general use of CPUE data is that for the Division 5 sperm whale stock there are extensive data from sightings by the Cheynes Beach spotter aircraft. These are considered to provide a more reliable measure of changing abundance.

When several nations take whales from the same stock, CPUE data can be obtained for each fleet and hence more than one estimate can be obtained for the initial population. These are then combined, usually by simple averaging, to give a 'best' estimate.

Estimates of the present number in each population component, and hence the present size of the exploitable population, are obtained by using the model with the best estimate of the initial population together with actual catch data. The same basic population model is then used to estimate the maximum sustainable yield for the stock and the population level at which it is available. This provides the information required to classify the stock and, if it is not protected, to set catch limits.

Criticisms of the scientific basis of management

The scientific basis for the present approach of the International Whaling Commission to whale management has three main targets for criticism: the assumptions and data on which the models are based; the applicability of these models to the problems of management when considered in relation to the ecosystem; and the rules defining the management procedure. These are considered below. More extensive discussions can be found in Holt (Inquiry Report, Volume 2), Project Jonah (1978), and FAO/ACMRR (1978).

Reliability of population estimates: There are still many aspects of the social biology of whales that are unknown. It was apparent to the Inquiry during the meeting of the Scientific Committee in Cambridge that this applied to certain of the biological assumptions made in the models and to the values of many of the parameters used.

For sperm whales the need for, or adequacy of, a reserve male pool of a particular size was one such question. The nature of the effect of a large scale reduction of males on the behaviour and breeding success of females was another. Both these factors can have a significant impact on the classification of a stock and the size of any available yield. Views expressed included on one hand a suggestion that socially mature males could move from harem to harem (fertilising each in turn) throughout the mating period, and on the other a hypothesis that these males stayed with a harem once contact was made, and that when the number of such males was reduced substantially some harems could fail to meet with a socially mature male.

Some authors (see, for example, Talbot (1975); Holt and Talbot (1978)) have pointed out more generally that the models do not make allowances for the impact of exploitation on behaviour or social structure, and the possible effect which the loss of group leaders, experienced in foraging or migration, may have on the survival of groups.

As there is substantial speculation on the extent and nature of social behaviour in whales, at the least it would be prudent to make a more conservative allowance for the possibility of effects of harvesting on social behaviour. For sperm whales in particular, until firm evidence is available on factors such as the behaviour of harem groups, the stability of their composition and their interaction with the 'surplus' socially mature males, it would be safer to assume that whaling has a disruptive effect which is not temporary.

It has earlier been mentioned that in modelling density dependence, the choice of the pregnancy rate response curve, and in particular the decision to use models with a maximum sustainable yield level at 60 per cent of the capacity of the environment, were essentially arbitrary although consistent with the best information available. The use of the 60 per cent level has frequently been said to be a conservative assumption, but as pointed out by Holt (Inquiry Report, Volume 2, pp.36-7) it could lead to larger catch limits than would be allowed if the true MSY level were at 50 per cent and this true value were used in the models. While setting the MSY level at more than 50 per cent is probably reasonable, in view of the uncertainty about the correct value it would again be prudent to use conservative catch limits.

The data generally used to indicate changes in whale density - that is, catch per unit effort (CPUE) data - have also been extensively criticised as not properly representing abundance changes. Corrections made for differences in sizes of catchers, weather conditions and other factors have been the subject of debate. There has been particular controversy about the adjustment to be made for the introduction of underwater scanning equipment (ASDIC), with estimates of the proper correction factor differing widely. While the Scientific Committee's present approach of using catch per searching hour where possible is likely to be an improvement, the Inquiry is not convinced that the reliability of the population estimates will be greatly enhanced. Many uncertainties remain about the assumptions made when adjusting the raw data. In any case it still leaves unresolved the question of whether statistics showing continuing high densities could be caused by whales grouping together when in fact there have been substantial reductions in overall numbers.

Fortunately the application of the sperm whale model in Division 5 is less susceptible to the main criticism of CPUE data since the abundance measures available from the spotter aircraft supporting the Albany operation do not require adjustment in the same way as the data from pelagic whaling fleets. They are considered more reliable because there has been less change in the procedures used. However, this has not prevented the Division 5 stock from being taken to levels where protection is needed.

Another assumption used in the models which is frequently said to be invalid is that the population before recent exploitation was stable and at the carrying capacity of its environment. This population is usually taken to be that of 1946 for sperm whales, although in some cases earlier data have been available and used. Best (1976(a)) examined the history of sperm whale catches and concluded that 'the balance of the evidence is that primitive whaling is unlikely to have significantly reduced most sperm whale stocks (one notable exception being the north Atlantic), and the assumption that 1946 stocks (particularly of mature females) were at their initial levels seems reasonable for most areas. The slow potential rate of recovery of the species, however, makes it essential that this preliminary finding should be more rigorously tested through more detailed analyses of primitive whaling records'. Holt (Inquiry Report, Volume 2, pp.44-5) comments on this question and it is further discussed in the extracts from the proceedings of the Scientific Consultation on Marine Mammals provided as Appendix 1 of his paper. These emphasise (as does Best) the importance of knowing what size and sex of whales were involved in these early catches, as the long-term effects on the stocks can vary substantially depending on whether whalers concentrated on females or solitary males. The implications of this for the population structure of the southern hemisphere stocks are discussed in Allen and Kirkwood (1977(d)). The assumption that a stock had restabilised at initial levels by a particular time would, if incorrect, lead to an optimistic view of the present state of that stock.

It is clearly impossible to determine statistically the accuracy of the estimates of the initial and current population sizes resulting from use of the models. An assessment of their reliability by several recognised experts is thus of some interest. Of course these views implicitly take into account deficiencies in the models beyond those already mentioned, including the considerations about the ecosystem discussed in the next section.

Dr Holt (Inquiry Report, Volume 2) reaches the general conclusion that current assessments of stock sizes are likely to be substantial overestimates. When asked to put a figure on this, Holt said: 'I should be surprised if the initial stock of any of the sperm whale Divisions was in fact less than half what is estimated by the Scientific Committee but I would not be very surprised if it was 20, 30 or 40 per cent, in fact, lower than has been estimated' (Inquiry Hearings, pp.475-6).

Dr Allen, Mr Bannister and Dr Kirkwood saw Holt's paper as a comprehensive and useful background survey of many of the issues facing the Inquiry. However, they did not support the impression given by it that 'population estimates are generally very inaccurate and also that they err in the direction of overestimation' (Inquiry Hearings, p.562). Kirkwood in relation to the Division 5 sperm whale stock said: 'I would be surprised if the true 1947 population (of exploitable males) was 14,000 as opposed to the estimate (of 18,200). I would be surprised if it was much above 20,000; it is very difficult to give an idea' (Inquiry Hearings, p.628).

The Australian scientists also considered that Holt's review of current harvesting strategies did not 'adequately recognise the progress which the Commission and the Scientific Committee have made in recent years in developing new approaches to the analysis of data and development of management practices' (Inquiry Hearings, p.562).

<u>Recognition of changes in the ecosystem</u>: The models have also been criticised for not taking account of changes in the environment. Oceanographic conditions can vary, and the frequent changes in abundance of many species within a marine ecosystem is amply illustrated by the

marked annual fluctuations in various fisheries. However, while whales might not naturally always maintain a steady population, it is generally accepted that they maintain a much greater degree of stability than species with a very high reproductive rate.[1]

Initial attempts have been made in the Scientific Committee to study the effect of random changes in the environment on the risk of extinction of stocks which are being exploited, but this work is not yet very advanced and needs to be taken further.

A more basic problem is that of long-term changes in the environment. The assumption of a density dependent response supposes that a population can potentially return to its initial level, and for a stock to produce a long-term sustainable yield it must always have a drive to return to this level. If, however, keeping the stock at a reduced level for a long period generates changes within the population or within the ecosystem so that the population tends to stabilise at the reduced level - for example if other species expand and take up the food available - then the yield estimated on the basis of the previous stable level is no longer relevant.

An example of this is the possible difficulties faced by protected baleen whale stocks as a result of the observed increases in both the potential population size for sei and minke whales and the abundance of crabeater and other seals and penguins. This was mentioned by Dr R.M. Laws, Director of the British Antarctic Survey and of the Sea Mammal Research Unit of the National Environment Research Council, in his introductory remarks in 1976 to the Scientific Consultation on Marine Mammals (FAO/ACMRR, 1978):

---

(1) These two types of animals are regarded as 'K-selected' and 'r-selected' species respectively. 'K-selected' species are good competitors, are slow to exploit new conditions and have low fecundity, slow maturation, long life, slow population growth, low energy expenditures for reproduction and a high level of parental care. The 'r-selected' species are not good competitors, are 'opportunistic' and have high fecundity, rapid maturation, short life spans and rapid population growth (FAO/ACMRR, 1978). (These classifications are of course descriptive rather than deterministic.) Marine mammals tend to be 'K-selected' species, and hence their unexploited populations could be assumed to be relatively stable.

A very important question is this: is the full recovery of previously overhunted, now protected species likely to be hindered by the interim expansion of other populations? What will be the eventual outcome in the Antarctic, for example, where the food available to other consumers has presumably increased concurrently with the decline in whale abundance and where a new man-induced perturbation - the development of a commercial krill fishery - is threatened, perhaps in the next ten to twenty years - who knows?

As Laws said: 'Clearly, we cannot sensibly study or manage marine mammal species or stocks in isolation but must take into account the interactions within the ecosystems of which they are parts'. Estimation of the long-term maximum sustainable yield for a stock would thus really require development of multi-species models and the incorporation of non-biological influences such as pollution. This degree of sophistication in ecosystem modelling for resource management seems far from realisation at this stage. In discussions which the Inquiry held with Dr Laws, he expressed the view that the single-species models currently used for stock assessment were the best available but stressed the urgent need for pursuit of advances in multi-species modelling. This would be important not only for the assessment of whale stocks but also to obtain a better understanding of the consequences of harvesting other marine resources, and particularly krill, which are rapidly becoming the subject of commercial interest.

The Australian Government has recognised the increased interest in the marine resources of Antarctica by extending its scientific programs into marine areas. The growing international interest in krill in the Antarctic was referred to in a speech on 4 October 1978 by the Minister for Science, Senator the Honourable J.J. Webster.

Referring to the fact that krill are largely restricted to the surface waters, the Minister said:
> They concentrate in dense swarms, often several hundred metres across, in which krill may be as concentrated as several kilograms in a cubic metre of water.

The estimated annual sustainable yield of krill is 100-200 million tonnes, compared with a total world fish catch of about 70 million tonnes.

Some scientists say krill's protein value is 20 times as great as that of beef, and that it could provide an almost everlasting source of nutrients for man and livestock.

Assessment of the commercial potential of krill is presently undertaken by USSR, Japan, Poland, Taiwan and West Germany.

The effect of a krill fishery on baleen whales, and in particular on the recovery of protected stocks, is of great concern to some scientists. Dr Holt at the public hearings referred to a recent meeting he had had with Dr R. Laws, Dr J. Beddington and Dr R. May to examine this matter. Their conclusion, admittedly using only the most general procedures, was that 'a relatively small krill fishery will drastically affect the recovery or productive capacity of the whales in the southern ocean'. He mentioned that in negotiations for a southern ocean treaty the general approach was that the krill population could be hammered quite hard without a very dramatic result, whereas his group believed that a krill fishery would have very significant and rapid effects on the ability of depleted baleen whale stocks to recover and of others to continue to give sustainable yields. The extent of the problem is shown by Holt's remark that if whaling now stopped and the Antarctic ecosystem were left alone it would 'take many centuries to re-establish anything like the original balance among species groups' (Inquiry Hearings, pp.465-7).

The Inquiry noted that the Scientific Committee was aware of the significance of such considerations. In fact in its assessment of the sei whale stock it introduced interspecies influences by allowing for changes in the density of blue and fin whales. Replacement yields - the catches which could be taken in a particular year without changing the stock size during that year - were used as the basis for setting catch limits for both sei and minke whales because their potential stock sizes appeared to be increasing. It is clearly accepted that the use of the maximum sustainable yield approach and the rules of the New Management Procedure are inappropriate in these circumstances.

Our concern with the importance of ecosystem considerations in the conservation of whales and with the adequacy of present models in this respect is well summed up in the words of Tranter:

> The test of whether the harvest has exceeded the capacity of populations to replace the harvest (by faster rate of growth and increased reproduction) is more immediate with short lived populations such as fish. In long lived populations such as whales the waiting time is relatively great in relation to man's rate of harvesting and capacity to deplete the stocks. In most cases the test has not yet been made. It is still proceeding. Several generations (of whales) must elapse before trends appear. The assumption is that, when whaling ceases on a stock, the numbers will eventually increase. This assumption is at best slightly suspect, at worst invalid.
>
> For recovery to take place, fecundity and survival must increase. This requires an increased per capita proportion of the resources available to the species, particularly food. Those resources are available not only to the threatened species but to every other animal in the community which uses the same food supply, and these animals may have the edge. They are smaller, have a faster rate of growth, greater fecundity; in short, a higher intrinsic capacity to increase. It is more likely that these will use the greater part of the resources available to the community of species of which the threatened whale stocks are part.
>
> The argument is speculative. There is no certainty. Only time will tell. The argument for reversability is equally speculative, perhaps more so. And again, only time will tell. If greater certainty is desired there is a ready option - stop the harvest. Even this may not be adequate - it could be necessary also to control man's exploitation of the food on which baleen whales browse (krill). This is another contemporary problem (Tranter, 1978, p.2).

<u>Management procedures</u>: As indicated in Chapter 3, the New Management Procedure of the International Whaling Commission classifies stocks by comparing their present level with an 'optimum' level. This is discussed further in the next chapter. Here we restrict ourselves to the current management procedure in which the MSY level is taken to be the optimum level - and which is therefore subject to the criticisms outlined above.

One general criticism of the procedure is its lack of flexibility. It assumes that it is possible clearly to identify separate stocks, and it virtually requires that each stock should be exploited and brought near to its maximum sustainable yield level. Holt (Inquiry Report, Volume 2, pp.17-19) points out that there may be advantages, for example, in differentiating between subsistence whaling, whaling for industrial purposes, and whaling for food for human consumption. If in some cases different management rules were applied to two stocks which were broadly comparable, this would also assist in obtaining a better understanding of the population dynamics of whales.

Lack of flexibility in the procedure was also referred to in personal submissions by Dr J.A. Gulland and Dr L.K. Boerema, scientists in the Food and Agriculture Organisation of the United Nations. Gulland considered that 'it seems clear beyond reasonable doubt that a sustainable yield can be taken from whales', but that there is a great deal of uncertainty about the size of the yield and the population level at which it occurs. He suggested that a more flexible approach be adopted in setting catch limits so that new information is taken into account without sharp changes in quotas. Dr Boerema expressed general agreement with these views, and in addition argued for a more flexible and more cautious management procedure which would take account of interspecies influences in the ecosystem.

Another criticism is that the procedure is based solely on the level of the population and does not take into account its composition by age and sex. This problem arises in the Division 5 sperm whale stock, where under present rules the most recent assessment indicates that males should be protected, but that a catch of females is available. This is despite the fact that the number of females would decline to protection status over the next decade even if no females were taken. It would clearly be better for the recovery of the stock for females also to be protected immediately.

The formula used to determine catch limits has also caused some concern, both for conservationists and the whaling industry. The problem is that when stocks are close to the maximum sustainable yield level, and particularly if they are slightly below this and verging on protection, small changes in assessments can lead to large changes in catch limits (see Figure 4.6, p.75). One example of this is the revision of north Pacific sperm whale quotas at the December 1977 Special Meeting of the International Whaling Commission in Tokyo. As a result of small changes in the estimated ratio of the present and initial population sizes, the catch limits were increased from the levels of 0 males and 763 females set at the June 1977 Commission meeting, to 5105 males and 1339 females. This ten-fold increase in the catch allowed has sometimes been wrongly interpreted as corresponding to a ten-fold change in the estimated stock size.

Following arguments that separate catch limits for the southern hemisphere Divisions led to problems for pelagic fleets moving throughout the Antarctic waters and reduced their efficiency, the Commission in 1976 adopted 'an arrangement whereby sperm whale catches of each sex in each Division might exceed the established quotas by not more than 10 per cent provided that the sum of the catches of each sex in all Divisions did not exceed the total quota for that sex' (Holt, Inquiry Report, Volume 2, p.29). Holt says that 'this provision is not likely to be a risky one, as far as the general state of the stocks is concerned', but points out that the allowance clause permits the maximum catch to reach 99 per cent of MSY, providing no safeguard against errors in assessments. The allowance provision, if used regularly to take additional whales from one stock (as has happened in Division 5), increases the risk of overexploitation.

The deficiencies in the methods for estimating MSY levels and catch limits and the criticisms of the New Management Procedure have been widely discussed in Scientific Committee meetings. The International Whaling Commission has endorsed the Committee's proposal for a working group to review the procedure, with the aim of developing a new approach to management which will retain the desirable attributes of the current procedure and avoid its shortcomings.

The Scientific Committee's concerns are summarised in Annex O of the 1978 Report of the Scientific Committee, prepared by Dr Allen as Chairman. This paper (IWC/30/4, Annex O) brings out many problems including:
    (a) 'Lack of a generally acceptable definition of optimum stock level.'

(b) 'Lack of evidence as to relation of the MSY level to the unexploited population level.'

(c) 'The requirement on Scientific Committee to make recommendations with a precision far in excess of the accuracy of the assessments on which they are based.'

(d) 'Frequent changes in assessments which have generally resulted from changes in the models and in the methods of analysis and not from real changes in stock sizes.'

(e) 'Large quota changes may result from quite small changes in the stock assessments due to the sensitivity of the quota-calculating procedure for stocks near MSY level.'

(f) 'The heavy load on Scientific Committee caused by the need to assess all stocks every year makes it difficult to devote sufficient time to critical analyses.'

(g) 'The system is designed to bring stocks to optimum levels as quickly as possible and therefore provides minimal opportunities to observe density dependent effects as stocks move from one level to another over a substantial range.'

(h) 'No specific provision for adjustments to quotas on basis of level of uncertainty of estimates.'

(i) Actual MSY stock level is almost impossible to define.

(j) Effects of past catches may invalidate the assumption used in the models that a stock is stable in some chosen year. This may mean that the replacement yield for a stock whose numbers are at the MSY level may in fact be less than the MSY and even the catch limits, causing the stock to decline.

(k) The concept of an MSY level is almost meaningless for stocks for which the carrying capacity of the environment is increasing.

The paper draws attention to the absence of ecosystem models or even multi-species whale models on which to draw for management purposes, mentioning that short-term adjustments to single-species models can improve current assessments but would not be adequate to forecast the future. It points out that any management procedure seeking to identify target population levels will face the same difficulty that has led to a somewhat arbitrary choice for the MSY level in the current procedure, but that procedures can be devised which move a stock towards a desired level without needing to know the numerical value of this level beforehand. While quotas should reflect stock size, with protection of stocks below a certain point, in any revised procedure, Dr Allen asserts, it may well be possible to relate quotas to a characteristic of the present population (such as a proportion of its replacement yield or of its gross biological production) which may reduce the dependence on sophisticated population models. His paper suggests that the abrupt quota changes inherent in the present procedure should be avoided and that it may be possible to develop quotas for more than one year at a time. In establishing the quota for a stock, the degree of uncertainty in its assessment should be considered.

As the summary above indicates, the Scientific Committee has taken account of many of the criticisms of the New Management Procedure's dependence on the use of the MSY level as a reference level, and has indicated possible alternative approaches to the current procedure. It is certainly important that the review group complete its task and its recommendations be considered promptly.

Conclusions

Our treatment of the management procedure and of the supporting modelling and assessment work of the Scientific Committee does not imply that the Inquiry would necessarily support exploitation if management and modelling were improved so that the criticisms mentioned were no longer relevant. It rather emphasises that in the context of international responsibility for the conservation of whales, the exploitation of whales by any nation must be managed with proper caution.

In examining the information available to it on the modelling and management procedures currently used, the Inquiry noted that some of the criticisms of the models would be better directed to the lack of information necessary to refine those models. There is a particular lack of information about the ecology and social behaviour of whales. However, we recognise that a perfect data base and management procedure is unattainable and some less perfect but practicable approach, refined as more information becomes available, has to be adopted.

The Inquiry is far from confident about the long-term recovery of whale stocks which have been drastically reduced and their ability to maintain their place in the ecosystem. The possible effects of proposals to harvest other marine resources, and especially krill, in the Antarctic are of particular concern in this respect.

For other stocks our general conclusions are in accord with the view expressed by Holt that the New Management Procedure 'has very greatly reduced, if not eliminated, the risk that any whale stock will be exterminated as a result of whaling' and that it has 'substantially reduced, though not eliminated, the delays in taking the necessary corrective actions' (Inquiry Report, Volume 2, p.71). However, there is certainly room for improvement in the management strategy.

If whaling continues, a more cautious approach is needed. At the moment we have only an inadequate understanding of marine ecosystems. Because of this, and because of the uncertainties in the procedures for assessing the current state of whale stocks and their available yields, it would seem prudent, as a minimum, to try to keep populations of whales relatively large until more conclusive evidence is available of their ability to recover from exploitation. If the International Whaling Commission had adopted a more prudent approach earlier in its history, whale stocks might now have been significantly larger. As it is, they have been excessively depleted.

5. INTERNATIONAL MANAGEMENT OF WHALES

The history of whaling and the development of controls on whaling have been discussed in Chapters 2 and 3. The International Whaling Commission has been widely accepted since 1948 as the international body responsible for the management of whaling.

This chapter looks at the future of international whale management and the attitude Australia should adopt towards the International Whaling Commission.

Management record of the Commission

Nearly all the submissions received by the Inquiry were critical of the past performance of the Commission.

The preamble to the International Convention for the Regulation of Whaling says in part:
> Recognising the interest of the nations of the world in safeguarding for future generations the great natural resources represented by the whale stocks.
>
> Considering that the history of whaling has seen over-fishing of one area after another and of one species of whale after another to such a degree that it is essential to protect all species of whales from further over-fishing.

The gravest indictment of the International Whaling Commission since its beginning is that it has presided over the decimation of blue and humpback whale stocks and the severe depletion of most fin and sei and some male sperm whale stocks. The current state of these and other stocks is discussed in Chapter 6.

The International Whaling Commission came into existence shortly after the end of the war, when there was a serious shortage of fats and oils.

The Commission was thus influenced from the start by urgent short-term economic considerations.

Whaling nations formed the International Whaling Commission as a result of their desire to stabilise whale oil production and world prices and concern for the financial stability of their whaling industries. In the 1940s and early 1950s the Australian Government, and almost certainly the Australian people along with the rest of the world, supported the Commission's priorities.

World support for, or indifference to, the Commission started to change as attitudes to the conservation of wildlife changed. In addition scientists started to issue strong warnings of over-fishing of some whale stocks and those warnings were ignored.

In 1960 the International Whaling Commission decided to ask three independent scientists to report on the condition of the southern hemisphere baleen whale stocks. In 1963 this group reported that, although not all analyses were completed, nevertheless the general conclusions, both qualitative and quantitative were clear, and pointed to the need for drastic and urgent action (IWC, 1964; Robinson and Waters, 1978).

The degree of concern felt by scientists was thus clear even in 1963 but no drastic action was taken for some years. There are a number of specific examples of the Commission repeatedly ignoring the warnings of its scientists until after the scientists' predictions had been proved correct.

The population of blue whales had been heavily reduced before the establishment of the Commission. However, although scientists started calling for special protection for the blue whale in 1953 and repeated the warning in subsequent years, extensive protection for the blue whale was not agreed to by the Commission until 1965, by which time there were so few blue whales left that protection involved no sacrifice by the whaling countries.

Scientists first advocated the abandonment of the 'blue whale unit' for setting quotas in 1963 (see Chapter 3). The use of the unit, a legacy from the Commission's earlier concern for stabilising oil production, resulted in the preferential hunting of the large blue and fin whales which led to blue whales becoming seriously depleted by 1963. Although humpback and blue whales where excluded from the unit following general protection, the notion of the blue whale unit was not abandoned until 1972, thus leading to over-exploitation of fin whales. The concept of management of individual stocks was similarly delayed, and its introduction for southern hemisphere sperm whales was not implemented in full accordance with scientific advice until 1976-77.

Since the Commission's beginnings considerable pressure has continually been applied by most of the whaling nations to maximise quotas, even in the face of scientific warnings of over-fishing. It was not until the period between 1962-63 and 1965-66 that quotas for the Antarctic were progressively reduced from 15,000 blue whale units to 4500 units, but by then the whaling countries were unable to catch the quotas in any year. Each of these years the quota was reduced to less than the actual catch for the previous year but each year there were not enough whales left to achieve the quota. By 1969-70 the quota had been further reduced to 2700 units.

A different example of the Commission's failure to ensure that its management was effective is provided by the delays in introducing the International Observer Scheme. An exchange of observers between whaling countries to help ensure that whalers complied with International Whaling Commission regulations was first discussed in 1955, and the scheme was discussed and deferred almost every year from then until agreement was reached in 1968. It was finally started in 1972.

There are thus major criticisms of the International Whaling Commission's stewardship of the world whale stocks up to 1972, but since then significant changes have been noticeable in the attitudes of the major whaling nations.

The 1972 meeting of the Commission was held two weeks after the United Nations Conference on the Human Environment had almost unanimously passed a resolution calling for a ten-year moratorium on commercial whaling. This clear expression of world opinion undoubtedly helped create the climate in which the Commission agreed on the following:

(a) the blue whale unit was finally abandoned and the Commission accepted the principle of setting separate catch quotas for all whale species, and for identifiable geographical stocks of the same species;

(b) separate quotas were set for male and female sperm whales in recognition of the different social behaviour of this species and of the preferential catching of males because of their greater size and value;

(c) agreements were signed between the whaling nations implementing the International Observer Scheme agreed to in 1968;

(d) it was decided to seek international support for a decade of intensified research on cetacea.

The new approach to management was embodied in the New Management Procedure introduced in 1974 (see Chapter 3). A key feature of this was that catch limits were finally linked directly with scientific advice, with the state of the stock being taken as the main consideration. The procedure enabled the Commission to place a selective moratorium on stocks according to their status, and in fact many stocks have since been classified as Protection Stocks (see Chapter 6 and Appendix 10).

In 1977 the International Whaling Commission passed two motions directed at encouraging those whaling countries which were not members of the Commission to join. Members were to take 'all necessary steps, including such amendments to their laws and regulations as may be required, to prevent the import into their countries of whale products from non-member nations'. The second resolution directed that member nations should take all practicable steps to prevent the transfer of whaling vessels and equipment, or the dissemination of whaling expertise, to nations that are not members of the International Whaling Commission.

The Inquiry also regards the acceptance of all except one of the recommendations of the Scientific Committee at the 1978 meeting of the Commission as a further indication of the changing attitudes of member countries. The one Scientific Committee recommendation not accepted at that meeting was the zero catch quota for the bowhead whale stock taken by Eskimo communities in the United States, and this change had the support of almost all of the conservation movements campaigning for an end to commercial whaling. It is also encouraging to note that no country has lodged an objection to a decision of the International Whaling Commission since 1973.

Problems in the management of whales internationally

Whichever organisation attempts the task of international management of whales, there will still be some unavoidable economic and political problems.

Most cetacea migrate extensively and, apart from some stocks of the smaller cetacea, could not be said to belong to any particular countries rather than to the world as a whole. There is thus pressure on each whaling nation to set substantial quotas and to take the largest possible share of these - generally to the detriment of the whale stock, other nations and even the long-term efficiency of its own industry.

When quotas were not divided between countries, the country with the largest whaling fleet could catch the largest share of the quota, and thus the whaling nations all increased the size of their fleets. In 1925 there were 23 factory ships attended by 234 catcher boats. In 1938 there were 37 factory ships attended by 362 boats. Many boats were destroyed during the war but in 1950 there were 26 factory ships attended by 468 catcher boats. All countries increased the number of catcher boats in an attempt to catch a greater share of the quota.

Quotas are now shared between countries according to agreed formulas and the problem takes a different form. Unless all the major whaling nations co-operate on whale management, it is not to the advantage of any individual nation to do so. Countries are unwilling to stop whaling because they believe that if they leave the whales others will take them. This argument was presented to the Inquiry to support the continuation of Australian whaling. Whaling countries are also given an advantage, at least in the short term, by any increase in the overall quota (or by the smallest possible reduction in the quota) and this is reflected in continual pressure to keep quotas at the highest possible level.

The economics of the industry must also be considered. Dr C. Clark has pointed out that the rate of natural increase for whale stocks is between 3 and 8 per cent which is less than the return available from other capital investments. His conclusion is that it is more profitable to exploit whale stocks extensively until there are so few whales remaining that it becomes too expensive to hunt them, when it is best to stop whaling and invest the profits in some other enterprise. The implication for whale management is that there will always be economic pressure to set quotas as high as possible, rather than accept the New Management Procedure which aims to stabilise whale stocks at around 60 per cent of their initial populations and set quotas just less than the estimated maximum sustainable yield of each stock (Clark, 1975). The economic pressure is greater because quotas have been falling steadily since 1962 and most countries have a number of whaling vessels lying idle which were built as part of the race to obtain the greatest share of the quota.

Another major difficulty facing any body managing whales is the nature of international organisations. Membership is voluntary and can be ended at any time. Decisions must be reached by consensus because a member country cannot be forced to accept a decision of an international organisation

As McHugh notes, international organisations 'have no magic powers of their own. They are simply creatures of the governments that established them, and they try to carry out the policies and wishes of the member nations and their people' (McHugh, 1974). The International Whaling Commission has failed to manage whale stocks effectively in the past because member countries could not agree on what action to take in response to the scientists' advice. For many years it was only the wishes of whaling nations that were reflected in Commission decisions - although any nation could have joined if it had chosen to - and the whaling nations, for many years, would not accept any management controls which they saw as threatening their whaling industries. It was not until whales began to be of interest to more than just those nations which caught them, and to be seen as important for more than just the products they provided, that any strong alternative view was expressed in the Commission.

The effectiveness of the International Whaling Commission has also been hampered by lack of money. Whichever organisation manages whaling will need to persuade the same countries that have been reluctant to give adequate finance to the Commission, to contribute more money towards the cost of administration and research. The position may well become worse unless more countries which do not themselves have a whaling industry are prepared to provide substantial funds.

Alternatives to the International Whaling Commission

With the above difficulties of international co-operation on whale management in mind, the Inquiry considered the suitability of some existing international organisations as alternatives to the International Whaling Commission. The topic was discussed at some length at the public hearings in Melbourne (Inquiry Hearing pp.539-557).

It is important that any alternative organisation be capable of managing the different aspects of the Commission's activities, including:
- (a) administration;
- (b) obtaining scientific advice on whale stocks;
- (c) establishing and implementing a management policy.

The main alternative bodies considered are discussed below.

United Nations Educational, Scientific and Cultural Organisation (UNESCO): UNESCO has not been involved in nature conservation and is not primarily a management body.

World Wildlife Fund (WWF): While WWF is concerned with wildlife conservation it does this by sponsoring other specialist organisations. The Fund has had no direct involvement with international regulation.

International Union for the Conservation of Nature and Natural Resources (IUCN): IUCN is an independent scientific organisation with one of the most comprehensive international memberships of any conservation body (including governments, government agencies and conservation organisations) and with considerable international prestige and scientific respect. Expertise on whales has been developed by IUCN through its Interim Committee on Marine Mammals and its Working Group on Management of Whales. The task of the latter committee has been to suggest more effective procedures for whale conservation and management. IUCN currently provides the secretariat for the Convention on International Trade in Endangered Species of Wild Fauna and Flora (CITES).

There are two important difficulties with IUCN as an alternative to the Commission. Governments may be unwilling to accept decisions of a body in which specialist government agencies and non-government bodies greatly outnumber national governments, and past policy statements by the IUCN on whaling may make it unacceptable to the whaling nations.

United Nations Food and Agriculture Organisation (FAO): FAO has the organisation and standing to manage international whaling. However its commitment to expanding world food production and the implicit acceptance that whales are a renewable resource that should be exploited for the benefit of man may make FAO unacceptable to countries arguing for the protection of whales.

United Nations Environment Program (UNEP): This organisation was the most suitable of those discussed. UNEP is committed neither to a policy of protection nor to exploitation of resources and as a United Nations body it would incorporate a degree of international accountability for the future of whale stocks that has been claimed as lacking in the International Whaling Commission. While UNEP has the organisation and resources to replace the International Whaling Commission it usually plays the role of catalyst and 'patron' of separate specialist organisations rather than assuming management responsibility itself. For example, IUCN secretariat functions for CITES are provided under a contract from UNEP. For UNEP to assume the role of the International Whaling Commission would require an extension of its present functions.

A proposal made by Dr Holt that the International Whaling Commission should continue to manage international whaling as an affilitated organisation of UNEP is worthy of serious consideration (Inquiry Hearings pp.545-551). This would seem to be a longer term development; in the meantime closer links between the Commission and UNEP should be encouraged.

There are thus no obvious immediate alternatives to the International Whaling Commission among existing international organisations. Further consideration would also need to be given to whether the charter of any such organisation provided adequate powers to implement the management recommendations. While the establishment of a new body could be considered, the advantage of the Commission is that it has existed for over thirty years, and thus possesses valuable institutional momentum. Membership of any international organisation must be voluntary, and the International Whaling Commission is more likely to retain the co-operation of whaling nations than any new body that is more directed to achieving protection. Problems that arise in reaching agreement in amendments to the Convention would not disappear with the establishment of a new organisation. Furthermore, any new organisation would need to draw heavily on much the same body of experts for scientific advice. Whaling nations can still exempt themselves from what is becoming more conservative scientific advice and it is not likely that this option would be surrendered in any new international organisation. Indeed, radical change within a new body could make it more difficult to retain membership of the major whaling nations.

The Inquiry has consequently concluded that there is no existing international organisation and little prospect of any new organisation being created that could be considered as a suitable alternative to the International Whaling Commission. A similar view has been put in many submissions. For example, IUCN (1978) said: 'criticisms notwithstanding, the Commission is the most effective existing organ of international whaling management and has the means for scientific and procedural improvement'.

The final question here then is, should Australia remain a member of the International Whaling Commission, and how would this decision be affected by any changes in government policy? Many conservation groups, including Project Jonah, the Australian Conservation Foundation and Friends of the Earth, have argued in their submissions to the Inquiry that Australia should continue its membership of the Commission. Noting the significant changes that have occurred in the Commission since 1972, the Inquiry on balance takes a similar view, so that whether Australia continues a policy of managed exploitation of whales or extends this to one of protection, the policy would be best pursued through continuing membership of the International Whaling Commission.

The future of whale management

There are some aspects of the Commission and its work that merit further examination. The Commission is itself working on these in reviewing its own charter, the International Convention for the Regulation of Whaling 1946. At its 29th meeting in June 1977 the Commission accepted an agreed negotiating text for revision of the Convention which was discussed at length at the preparatory meeting held in Copenhagen in July 1978.

One major issue is whether an international body broadly responsible for the conservation of all cetacea is needed. While this would have to recognise the rights of coastal states in relation to stocks remaining largely in national waters, there would be advantages in a co-ordinated international approach for all species. In particular, proposals such as that of Canada of asking the Commission's Scientific Committee to be available as a central body for scientific advice on the management of all cetacea would seem to merit further consideration.

The Commission has often been criticised for being too much directed to the exploitation of whales. One example of this is that under the present management procedure, as an interim measure the 'optimum' level for a stock is taken to be the maximum sustainable yield level. This does not take into account the many other objectives which individuals or groups in society may have in relation to whales, which may be as diverse as obtaining pleasure from watching whales, reducing inhumanity towards animals or maintaining stability of the ecosystem.[1]

International concern that the Commission should take a broader approach to management of whales, or indeed all cetacea, is reflected in the amendments to the Convention which have been suggested. For example, the United States has proposed that the title of the Convention be 'International Convention

---

(1) Objectives pertinent to whale management were discussed at the Bergen Scientific Consultation on Marine Mammals (FAO/ACMRR, 1978) and a list of these is provided as Appendix 9.

for the Conservation of Cetaceans', and Canada has suggested that the
preamble to the Convention should include the words 'recognising that
whales/cetaceans may be rationally utilised on the basis of their
nutritional, economic, social and aesthetic values' (IWC/29/24, 1977).

It has also been proposed that the Convention should require the
Commission to establish and maintain close relationships with other
inter-governmental organisations which have interests in cetacean conservation
including particularly UNEP, FAO, CITES, an appropriate body in relation to
the proposed Convention for Conservation of Antarctic Marine Living Resources,
and certain other regional and specialised fishery bodies.

The Inquiry considers that these proposed changes to the International
Whaling Commission and its objectives and responsibilities would assist
in improving the conservation and management of all cetacea internationally,
and accordingly that they should be given support by Australia during its
continuing membership of the Commission.

6. WHALE STOCKS AND WHALE RESEARCH

Each stock of whales has attracted a different degree of commercial interest. This chapter briefly reviews the status of the major great whale stocks before turning more specifically to those species which have been hunted in the waters around Australia - sperm, humpback and right whales. Brief reference is also made to small cetacea off Australia.[1] The list of cetacean species in Appendix 5 indicates those found in the waters around Australia.

The extent and direction of future research in Australia is also considered in the light of the information needed to conserve whale stocks.

General

The Schedule to the International Convention for the Regulation of Whaling is amended annually to give the current stock classifications and consequent catch limits or prohibitions imposed on each stock. Table 1 in Appendix 10 provides a summary of these provisions for the 1978-79 Antarctic season and the 1979 season for areas outside the Antarctic, and includes estimates for each stock of its initial and current sizes.[2]

Some generalisations from the table are obvious. The stocks of blue, humpback, gray, right and bowhead whales have been protected for long periods because they were reduced to very low levels by exploitation, although some, such as the pygmy blue whale and the western Atlantic humpback whale,

---

(1) No attempt has been made in this report to present a comprehensive assessment of stocks of small cetacea, because of their relative freedom from interference by man and the general lack of knowledge about current stock sizes.

(2) The use of one decimal place in Table 1 in Appendix 10 should not be regarded as an indication of accuracy. In fact many entries are 'best estimates' from a range of figures or are single estimates from rather scant data. The reliability of estimates such as these has been discussed in Chapter 4.

were reduced less than others. The only real recovery has been that of the eastern Pacific (California) gray whale stock, while, in contrast, the western Pacific stock has possibly become extinct. There is also encouraging evidence of a significant rate of increase in right whales off South Africa, although the numbers remain small. However, the overriding observation for these five species of whales must still be that a total of something in the order of 450.000 whales of an exploitable size has been reduced to less than one-tenth of that number.

Turning to the more recently protected stocks it is apparent that most fin and sei whale stocks are now protected because they have been over-exploited and that several sperm whale stocks are or may soon be in this position. Generally the protection granted in these more recent cases is less an indication that stocks are in danger of extinction than it was for earlier ones. However, an examination of the reduction in the exploitable population for the stocks involved shows once again that it is substantial. Their numbers, very approximately, have been reduced from perhaps 600,000 to perhaps 150,000. As a gross comparison, a stock of 600,000 whales would be held at 360,000 under the current New Management Procedure with an MSY level at 60 per cent of initial size. While in general the reduction of fin and sei whale stocks has been indeed substantial, the Area VI fin whale stock, the Area II sei whale stock, and some of the north Atlantic stocks seem to be less affected.

The main species exploited at present are sperm, Bryde's and minke whales. The estimates given in Table 1 of Appendix 10 for the various sperm whale stocks must be considered in terms of the criticisms raised in Chapter 4. In particular, the figures shown for southern hemisphere Divisions other than Divisions 5 and 6 take no account of the change to 'catch per searching hour' as the preferred indicator of abundance. This may have a substantial effect on the estimates. Even then, there are some who would consider that the estimates are still likely to be optimistic (Holt, Inquiry Report, Volume 2, pp.72-3).

On the current estimates it would seem that female sperm whales have, in the most heavily exploited stocks, been reduced to levels of about 55 to 60 per cent of initial levels. However the majority of female stocks are still at more than 85 per cent of their initial level. In contrast, most male stocks have been reduced to below 50 per cent of their initial levels, with the Division 5 stock the lowest at 26 per cent. Even though sperm whales are polygynous, the poor understanding of relevant aspects of the social behaviour of the sperm whale leaves open to conjecture whether the stocks can cope with such cuts.

The status of the stocks of minke and Bryde's whales, which are the remaining species of commercial significance, is not known accurately but in general does not give cause for concern. Like sei stocks, minke stocks appear to have been increasing before they began to be exploited, probably benefiting from the reduced competition for food from the larger baleen species. Current quotas are equivalent to the estimated net annual increase in stock size (the replacement yield). In theory this should maintain the stocks at their current level, in effect restraining what might have been their increase at the expense of the larger baleen whales. However this may simply allow greater increases of species such as the crabeater seal which also feed on krill. The expanding fishery for krill may also reduce the minke stocks' potential to increase (and prevent them from supporting the current rate of harvest).

The exploitation of minke stocks has not been without some problems. Apparently as a result of the distribution pattern of the species, catches have tended to include a larger proportion of females than males.[1] To compensate, current limits on southern hemisphere catches have been based on estimates of the replacement yield for females, the male limits being calculated on the basis of their proportion in the catch. While this should overcome any excessive reduction in numbers of females relative to males, exploitation will still take sexes disproportionately.

Bryde's whales have probably not been subject to the interest of pelagic fleets to the same extent as other species because they penetrate less into higher latitudes. Despite this, they have been taken by land-station operations and, because a decline has been observed in the average length of females off South America, the Commission has requested that

---

(1) Male baleen whales, unlike male sperm whales, do not grow larger than females, so gunners are usually unable to distinguish between the sexes.

Peru should not increase its level of catch until further studies have been carried out. In general the status of Bryde's whale stocks is not known; the only stock on which the Commission allows a commercial take is the western north Pacific stock for which a reasonable assessment has been developed. Other stocks, while regarded as at or near initial levels, are protected until a satisfactory population assessment has been developed. An exception to this has been the issue of scientific permits by Japan in the last three years for substantial takes of Bryde's whale for scientific purposes.

Sperm whales off Australia

Sperm whales in the waters around Australia are regarded as representing two stocks, a western stock (the 'East Indian Ocean' stock) and an eastern stock (the 'East Australian' stock). These correspond to Divisions 5 and 6 of the International Whaling Commission (see Chapter 1 and Appendix 6).

Table 6.1 shows catches of male and female sperm whales by Division in the southern hemisphere from 1946, when recent intensive exploitation began, to 1977. It also includes the annual catches by Australia in Division 5 over this period. Up to 1977 these totalled 14,160 whales (approximately 11,800 of which were males) which were all taken from Division 5 except for one or two individuals on rare occasions from Division 6. The total catch by all nations over this period was approximately 29,800 from Division 5 and 13,700 from Division 6. Table 2 in Appendix 10 provides further details for individual countries of southern hemisphere sperm whale catches and catch quotas for the period 1973 to 1977.

Since the establishment of catch limits, the USSR and Japan have always taken the major portion of the sperm whale allocation for the southern hemisphere. The remainder has been shared between land-stations with limited access to stocks. Australia negotiated to obtain the basic Division 5 catch limits for the 1977 and 1978 seasons, but the pelagic fleets took some whales within the Division in this period by using a large proportion of the 10 per cent allowance provision. (The risks involved in this provision were discussed in Chapter 4.) The Division 6

Table 6.1
Southern hemisphere sperm whale catches by Division,[a] 1947-77, including catches by Australia

	Division 1			Division 2			Division 3			Division 4			Division 5		
	Males	Females	Total[b]	Males	Females	Total[b]	Males	Females	Total[b]	Males	Females	Total[b]	Males	Females	Total[b]
1946-47[c]	184	0	184	220	2	222	736	206	942	413	0	413	214	0	214
48	135	0	135	501	19	522	1082	442	1524	806	0	806	416	0	416
49	364	0	364	980	90	1073	1928	241	2168	412	0	412	313	0	313
50	221	0	221	935	46	983	1180	84	1265	332	0	332	311	0	311
51	373	0	373	1244	169	1413	1826	267	2093	378	0	378	980	0	980
52	308	0	308	1503	108	1611	1615	33	1648	730	0	730	519	0	519
53	258	0	258	925	63	994	1014	43	1057	241	0	241	108	0	108
54	370	0	370	876	0	876	1137	41	1178	198	0	198	176	0	176
55	172	0	172	1496	0	1496	2514	235	2749	675	0	675	587	1	588
56	282	1	283	1391	0	1391	1752	137	1889	1017	0	1017	1064	0	1064
57	354	2	356	1514	142	1669	1159	282	1447	140	0	140	443	17	460
58	435	0	435	2001	294	2298	2024	268	2295	559	0	559	997	37	1034
59	320	3	323	1637	252	1889	1694	281	1982	930	0	930	812	0	812
60	344	18	362	1239	221	1460	1528	376	1911	479	0	479	1018	10	1028
61	560	46	606	1623	340	1972	1433	444	1887	364	0	364	1005	3	1008
62	592	28	620	2598	264	2873	1817	695	2536	667	7	674	773	14	787
63	732	52	786	1967	318	2285	1872	1087	2959	591	203	794	1027	59	1086
64	460	37	497	1923	156	2079	2317	1204	3521	1814	927	2741	1334	124	1458
65	1182	10	1192	1193	257	1452	1651	1810	3461	1448	635	2083	2009	129	2138
66	497	28	525	1366	163	1530	1751	1093	2844	508	17	525	1683	198	1881
67	304	11	315	1117	285	1402	1748	690	2438	1053	88	1141	1153	68	1221
68	38	32	70	580	25	605	1632	247	1879	727	19	746	850	93	943
69	1	0	76	791	61	852	1456	965	2421	565	36	601	940	48	988
70	36	46	82	1129	86	1215	2041	975	3016	1190	72	1262	1154	51	1205
71	525	63	588	991	53	1044	1850	1103	2953	894	322	1216	1088	114	1202
72	1257	149	1406	1098	82	1180	2812	692	3504	549	31	580	1096	380	1476
73	1161	855	2016	349	848	1197	2118	885	3003	525	62	587	1022	287	1309
74	923	823	1746	112	0	112	1166	921	2087	219	11	230	1069	493	1562
75	1214	1303	2517	8	0	8	977	810	1787	253	243	496	934	561	1495
76	1249	994	2243	388	424	812	406	0	406	417	371	788	942	410	1352
77	309	74	383	785	194	979	632	183	815	590	0	590	555	139	694
													26592	3236	29828

Division 5 total:

Year	Division 6			Division 7			Division 8			Division 9			Australian catch		
	Males	Females	Total(b)	Males	Females	Total(b)	Males	Females	Total(b)	Males	Females	Total(b)	Males	Females	Total(b)
1946–47(c)	4	0	4	0	0	9	0	0	0	2993	563	3607	0	0	0
48	145	0	145	0	0	0	0	0	0	2561	617	3228	0	0	0
49	258	0	258	346	0	346	34	0	34	387	293	731	0	0	0
50	139	0	139	52	0	52	23	0	23	628	141	769	0	0	0
51	469	0	469	458	0	458	52	0	52	5931	1119	7098	0	0	0
52	628	0	628	627	0	627	84	0	84	148	0	828	0	0	0
53	79	0	79	147	0	147	31	0	31	369	225	1968	0	0	0
54	103	0	103	209	0	209	156	0	156	396	226	4932	0	0	0
55	277	0	277	240	0	240	157	0	157	284	158	2615	6	1	7
56	592	0	592	557	0	557	340	0	340	1670	1385	3534	61	0	61
57	170	0	170	559	0	559	433	0	433	3049	2189	5238	122	17	139
58	325	0	325	589	0	589	497	0	497	2792	2031	4824	247	36	283
59	299	0	299	396	0	396	639	0	639	2942	2263	5471	138	0	138
60	703	6	709	102	0	102	88	0	88	273	38	5519	273	9	282
61	373	0	373	697	0	697	41	0	41	1271	983	5928	452	3	455
62	205	0	205	156	4	160	64	0	64	1274	1134	5712	577	16	593
63	670	175	845	164	0	164	154	0	154	2663	2288	4951	599	49	648
64	1101	99	1200	1506	1073	2586	0	0	0	614	431	3242	716	85	801
65	1517	1	1518	1013	1120	2133	83	18	101	904	466	1370	647	105	752
66	646	242	888	1125	262	1387	247	0	247	335	142	1542	595	11	606
67	301	9	310	1755	24	1779	432	0	432	229	46	1036	560	26	586
68	395	43	438	987	14	1001	72	0	72	776	821	1932	583	75	658
69	257	10	267	927	90	1017	234	0	234	1094	358	1707	636	43	679
70	500	26	526	586	140	726	50	0	50	953	483	1706	776	23	805
71	210	50	260	1439	1281	2720	58	0	58	841	480	1567	820	40	864
72	260	144	404	810	323	1133	79	0	79	691	859	2078	792	161	955
73	575	249	824	772	1172	1944	338	25	363	614	881	1879	684	287	971
74	361	502	863	1571	1598	3169	603	172	775	539	747	1286	629	450	1081
75	274	108	382	946	1034	1980	1679	442	2121	556	237	793	692	480	1174
76	0	0	0	455	328	783	731	839	1570	912	588	1500	650	345	997
77	213	3	216	0	94	94	745	208	953	316	197	513	508	116	625
Division 6 total:	12,049	1667	13,716										11,763	2378	14,160

Australian total:

Footnotes to Table 6.1

(a) Excludes pelagic catches north of 40°S before 1962-63.
(b) Includes whales lost and those of unknown sex (especially of significance in Division 9).
(c) 1946-47 pelagic and 1947 land-station season; similarly in subsequent years.

Source: 1946-73 - From Table 2B, Best (1976(a)).
1973-74 and 1974-75 - Extracted from BIWS 10° Square tables; G. Kirkwood (pers. comm.).
1975-76 and 1976-77 - Bureau of International Whaling Statistics, Sandefjord.
Australian data - 1947-61 and 1969-77 from Commonwealth Government (1978) Tables 2 and 6; 1962-68 from Bannister (1974) p.241 to include research catches (1958 and 1961-63 further modified according to Bannister (pers. comm.) to include Carnarvon commercial and research catches).

stock is exploited solely by the pelagic fleets and catches have mostly been far below catch limits.

At the 1978 meeting of the Scientific Committee it was intended to make new assessments of the southern hemisphere sperm whale stocks using the methodology outlined in Chapter 4 and Appendix 8. However the Committee was not able to complete its assessments at that meeting and suggested adoption of interim catch limits based on the limits for the 1977-78 and 1978 season and taking account of the one assessment it had been able to complete for Division 6. This assessment used the Japanese 'catch per unit effort' (CPUE) data which were believed to be representative for that Division. The International Whaling Commission reduced the limit on males for Division 5 by 25 per cent from 536 to 402 for the 1978-79 and 1979 season, retaining the Sustained Management Stock classification. The limit for females was reduced by 10 per cent from 177 to 159, retaining the Initial Management Stock classification. For Division 6 both males and female stocks retained Initial Management Stock status, the male limit remaining at 276 and the female limit being reduced by 10 per cent to 83. The 10 per cent allowance was removed for all Divisions.

In the subsequent international negotiations Australia again obtained the full Division 5 quota.

Division 5: The most recent assessment of the status of sperm whales in Division 5 was submitted to the Inquiry in August 1978 by Dr Kirkwood, Dr Allen and Mr Bannister. This is included as Appendix 11. It brings out the consequences of the level of exploitation of the Division 5 stock since 1946. Of particular significance is that the average annual catch from 1946 to 1977 of 857 male sperm whales has been far above the estimated maximum sustainable yield of 464. The male population is now at 26 per cent of the initial population level and so is 19 per cent below the maximum sustainable yield level (which is at 32 per cent of the initial population level). Hence, applying the New Management Procedure, it should now be classified as a Protection Stock, with a catch limit of zero. The model indicates that even in the absence of catches the female stock would also fall to that category by about 1989. While the New Management Procedure does not formally recognise what will happen to a stock in the future and technically would allow a current take of about 131 females, the scientists consider there is good reason to stop taking females from this stock immediately as this would permit the most rapid recovery of the stock.

The stock has been reduced to this condition because of an excessive reduction in the number of socially mature males. In a relatively few years time when the ratio of males to females readjusts, some males could again be taken every year, although, as the scientists say, the factors affecting the safety of this catch would be complex and would require careful monitoring. In contrast, the female stock will take 35 to 40 years to reach the maximum sustainable yield level again.

While this assessment is the most recent available and tends to support the attitude of many submissions that the stock condition is far from satisfactory, it must be noted that it is subject to most of the criticisms discussed in Chapter 4. However, the Inquiry is satisfied that, even if details of the results may be debated, they are sufficiently reliable to justify an urgent revision of the stock classification and catch limits set for the Division by the International Whaling Commission in June 1978. The abundance indicators from the spotter aircraft are regarded as more reliable than indicators from the land-station CPUE or pelagic fleet CPUE data. The decline in pregnancy rate seems consistent with the decline in the number of males. Spotter plane sightings of males for April to July 1978 are consistent with sightings for previous seasons (Bannister, letter to the Inquiry, 29 September 1978), and so give no cause to believe that their inclusion would change the above conclusions substantially.

It is the opinion of the Inquiry that the need for caution in management practices as discussed in Chapter 4 would make any early resumption of whaling most imprudent, and accordingly both male and female sperm whales in Division 5 should be given extended protection.

Giving full protection to the Division 5 stock for an extended period would also have the considerable merit of allowing an investigation in more detail of the capacity of a previously exploited stock to re-establish its initial stock level. At the same time, there may also be advantages in extending protection until the consequences of the developing interest in exploitation of various southern ocean living resources associated with the whales' ecosystem are clarified.

What course should Australia then take within the International Whaling Commission in relation to the quota which was set for the Division 5 stock at the June meeting? That decision was made by the Commission on the best advice that could be provided at that time by the Scientific Committee. Appropriate steps have already been taken by the Australian Government to have the matter of the present state of the stock raised at the special meetings of the Scientific Committee and the Commission to be held in November and December 1978. If, having regard to the evidence recently provided, the Scientific Committee and the International Whaling Commission in turn consider that a different view would have been taken in relation to that stock if the present evidence had been placed before it in June, then the Commission might reasonably be expected to take account, by some appropriate procedure, of the new evidence.

But what would be the position if in spite of the Australian analysis the quota were left unaffected? In the past it is true that in several cases International Whaling Commission catch limits have not been taken up. We refer to instances over the years in which quotas allotted to the Netherlands, Norway and South Africa were involved. In each of these cases the quotas were taken up by other governments. The case of Canada has also often been referred to, but this case is different. The stocks previously exploited by Canada are either restricted to land-based operations under the Schedule to the Convention or are available in numbers too small to be attractive to the pelagic fleets. We are not aware of any country wishing to take whales from these stocks since Canadian whaling stopped.

There is however a new factor in the present situation. It is the power of a coastal state to declare a 200-mile fishing zone off its coast and to prohibit whaling in that zone as has already been done by the United States and New Zealand. This is important because of the migration path of sperm whales off Australia. For much of the year large groups of females, their young and juveniles are found in the waters near the edge of the continental shelf, which in the case of Western Australia is well within the 200-mile fishing zone. Male groups roam over wide areas of the ocean but they are naturally also found for substantial periods of the year close to the continental shelf. Accordingly similar legislation by Australia would appear very likely to secure a considerable degree of protection for the sperm whale stock found in Division 5. (Some protection for the Division 6 stock would also be secured within the zone along the east coast of Australia.) No doubt any government which displayed an interest in a quota for the Division 5 stock would take into account the rights of the Australian Government in relation to its 200-mile fishing zone.

Division 6: No formal account was available of the single assessment of Division 6 mentioned previously. Dr Kirkwood, who carried out the assessment at Cambridge, provided the following advice on 4 September 1978 in a letter to the Inquiry:

> The previous assessment put the male stock at the beginning of the 1977-78 season at 45.1 per cent of initial (Initial Management Stock) with a catch limit of 276, and females at 93.5 per cent (Initial Management) with catch limit 92. The new assessment...put males at 40.2 per cent (Sustained Management) with catch limit 247 and females at 91.8 per cent (Initial Management) with catch limit 70. The Commission agreed upon catch limits, based on the previous assessment, of 276 males and 83 females...
>
> You will be aware that these are at best tentative figures, subject to revision at the future Scientific Committee meetings. However they indicate that the quotas in Division 6 on proper reassessment will be below the 1976-77 quotas, but that in all likelihood both males and females will be estimated as being above MSY levels. Thus it is most likely that under the current schemes, whaling will continue in Division 6.

In the light of the discussion in Chapter 4 of the accuracy of assessments such as this, it would seem prudent for the International Whaling Commission to take a more conservative approach in establishing catch limits for this and other sperm whale stocks where whaling may continue.

Humpback whales off Australia

Dr R.G. Chittleborough, formerly responsible for CSIRO's research on local stocks of humpback whales, prepared a detailed review for the Inquiry of Australian humpback whaling from the beginning of the century until its end in 1963. His paper, which also provides an insight into the management of humpback whale stocks since 1949, is provided as Appendix 12.

Chittleborough reports that the original size of the Area IV[1] (Western Australian) population before any hunting had taken place was in the region of 12,000-17,000 individuals, and that in 1949 the stock contained approximately 10,000 individuals. Upon the information now available, this is the level at which the population would have had its maximum rate of increase which would have been approximately 390 whales per year. The catch statistics however show that for thirteen consecutive years (1950-62) the total catch from this population exceeded that level, so that signs of depletion soon became apparent. When the International Whaling Commission decided in 1963 that humpback whaling should stop in the southern hemisphere, the population 'had been reduced to less than 800 individuals, more than half of which were immature'.

Chittleborough indicates that the original size of the Area V (East Australian-New Zealand) population of humpback whales was somewhat less than that of the Area IV population, containing approximately 10,000 individuals, and up to 1950 this population was close to its original size. A population of about 8000 would have given the maximum rate of increase of approximately 330 per year. Catches were much larger than this and thus a major depletion of this stock had to follow. When the fishery stopped in 1962 the remnant of this stock was estimated to be close to 500 whales.

Chittleborough's conclusion on the state of these stocks is that provided they 'were not depleted further from those few remaining in 1963, they should not be in danger of extinction. As this species migrates so close to the shore during the breeding season, prospects for successful

---

(1) The baleen whale Areas are described in Chapter 1 and Appendix 6.

mating and care of young appear to be better than for species which are scattered more widely in oceanic waters during their breeding season... However, owing to the present low levels of these populations and the relatively small net rate of increase, it may take up to 35 years for the Area IV population...and up to 50 years for the Area V population to recover' to the maximum sustainable yield levels.

A summary of recent observations of the Western Australian stock of humpback whales was provided by Mr J.L. Bannister of the Western Australian Museum in his submission:

> There has thus been growing interest recently in the possibility that humpback stocks might be showing early signs of recovery, and one attempt to monitor the situation has been in operation off Western Australia since 1976. Others, off the east coast, have also begun recently or are contemplated. Off Carnarvon, a 'box search' of flights 10 miles apart up to 80 miles from shore - covering the main northern migratory path as recorded off the west coast earlier - was undertaken in 1976, once a month in June, July and August.
>
> Despite poor weather conditions in one month, enough of the proposed flight plan was completed to show that humpbacks were present as expected, off Carnarvon, but indices of abundance could not be compared with those from the last year of commercial operations there (in 1963) because of the relatively small amount of flying time in 1976. A revised program was therefore devised for 1977, restricting the flying time to one month (July) when whales could be expected to be moving northwards, and when weather conditions were likely to be good. In 7 flights out of Carnarvon in 1977, 30 sightings of humpbacks were recorded. While the amount of time spent flying was still considerably less than that for which data are available from 1963 it was considerably more than for 1976 in the same month (see Table 6.2). For strict comparison between 1963 and present day results, the most appropriate index would be 'whales seen per flying day' since the commercial spotter flights in 1963 presumably spent some time searching in a particular area, but would only have logged any sightings once; on present flights the flight path is covered only once per flight. On the other hand, the best comparison between 1976 and 1977 data should be between

whales seen per flying (or searching) hour, but the 1976 figures are so small as to make such comparison very difficult.

Table 6.2
Aircraft sightings of humpback whales, Carnarvon area
July 1963, 1976 and 1977

	1963	1976	1977
Number sighted	120(110)[a]	9	30
Days flown	20	1	7
Hours flown	126.54	5.17	25.00
Whales per flying day	6.0	9.0	4.3
Whales per flying hour	0.87	1.74	1.20

(a) Figure in brackets represents numbers seen for which flying hour data are available

With the flight plan repeated in future years, however, it should be possible to obtain more comparable data for the same period each year.

None the less, from these data it seems most unlikely that there are fewer humpbacks off Carnarvon, at least in July, than there were in 1963.

Other reports from the Carnarvon area and elsewhere along the coast seem to confirm this conclusion. One report, from conversations with local fishermen in Shark Bay, was of a peak in numbers at the end of June (1977) when for several days 30 to 40 humpback whales on average were counted every day travelling north (K. Godfrey, pers. comm.) (Bannister, 1978, pp.12-14).

Recent studies of the eastern Australian humpback stock provide some indication of its current status. Dr M.M. Bryden (1978) of the University of Queensland has suggested that on the limited evidence available, the Area V population of humpback whales is still extremely small and may even have decreased since protection in 1963. His observations are based on limited aerial and shore-based searches in the Moreton Island to Byron Bay region. Also Dr W. Dawbin, who recently commenced a sighting program off the coast of New South Wales and Tasmania, has described the results so far as disappointing and disturbing (Inquiry Hearings, pp.273, 532).

However, Dr R. Paterson, who has had a long interest in the annual migration of these whales, has observed an apparent increase in the numbers of the small residual stock passing by Point Lookout on Stradbroke Island where his family has had a house since 1948. In the period 1968 to 1973 he did not sight a whale from Point Lookout (although he pointed out that his time there had been limited), but from 1974 onwards he has seen a steady reappearance of whale groups.

His submission reported 12 sightings of humpbacks on seven days (including 4 days on which there were no sightings) during the 1978 June and July northward migration, and no sightings on 10 days during the August to October southward migration. Reliable casual observers at Point Lookout on different days reported another 15 sightings during June and July and 28 sightings (including 8 sightings at Cape Byron 150 km south) during September and October (Paterson, 1978).

Whether Dr Paterson's sightings provide evidence of some recovery in the stocks must await further observations; his sightings have to be considered in the context of the wide area covered by the Area V population (from the Great Australian Bight to beyond New Zealand). However, as the coastal waters in the vicinity of Cape Byron seem to be a major pathway for migration of the Area V stock, the small numbers still give cause for concern about its current numbers.

Right whales off Australia

As indicated earlier, right whales were common in the bays of southern Australia at the time of first settlement and were soon the object of intensive bay whaling operations across the whole southern part of the continent. The slow swimming whales were easy prey and their numbers were reduced drastically. Right whales were the first species to be accorded total protection when international agreements on the management of whaling were developed.

The Inquiry received a detailed account by Mr Bannister of recent observations of the species in Western Australian waters:

Before 1955 there were no reported southern right whale[1] sightings off Western Australia (Chittleborough, 1956) but since then sightings have been recorded fairly regularly, mostly in late winter or spring when individuals, particularly females about to calve, come close inshore. Following reports of a recovery in right whale numbers off South Africa (For example, see Best, 1970) a series of spotting flights was begun, between Cape Leeuwin and Israelite Bay, along some 900 km of the southern W.A. coastline, to record the presence of right whales there from August to October.

While poor weather and unavailability of aircraft and observers combined to reduce the amount of flying actually carried out, enough was achieved in 1976 to show that at that time of the year only very small numbers of this species were close to the southern Western Australian coast, and certainly not approaching the numbers reported off South Africa recently, where over 100 were seen on one flight along a comparable length of coastline (Best, 1976 (b)).

In view of reports of sightings in other months, the flight plan was extended to cover the period July to November/December in 1977. In that year 21 confirmed sightings were made - considerably more than the 3 seen in the previous year - with a peak of sightings in September, and with a peak then also of females reported with calves (Bannister, 1978, p.14).

The numbers reported are thus extremely small.

Presumably because of the extremely low numbers of this species, there was little information available to the Inquiry on the status of right whales off the south or east coast of Australia. The only data were details of strandings in Tasmania, and some occasional reports of sightings by fishermen, provided by Dr E.R. Guiler, Reader in Zoology at the University of Tasmania, in a letter to the Inquiry on 11 September 1978.

---

(1) The southern right whale is sometimes considered a separate species within the genus Eubalaena (see Appendix 5).

Small cetacea

Many of the small cetacea recorded in waters around Australia are known only from rare instances of strandings; we set these aside. Very little information was available to the Inquiry even on the small cetacea more commonly found. Most studies of them have been concerned with classifying and describing them, or have been concerned with maintaining the animals in a healthy state in captivity.

There are some indications that very small numbers of the animals have been taken or killed deliberately for fishing bait or to drive off schools of dolphins dispersing schooling fish which were the object of a commercial fishery. Few details are available. Also, instances have been mentioned where dugongs (Dugong dugon) and small cetacea (including the Irawaddy dolphin Orcaella brevirostris) are caught incidentally in the course of gill net fishing operations in northern Queensland (Heinsohn, 1978). In some parts of the world very large numbers of dolphins have been killed in purse-seine fisheries but these problems have not been encountered in Australia as different species and different methods are involved.

A more significant but still quite limited take of animals occurs unintentionally in the course of mesh netting operations aimed at capturing large sharks near ocean bathing beaches. This is a regular activity close to the popular surf beach resorts near the main population centres of New South Wales and Queensland.

Dr R. Paterson provided the Inquiry with details obtained from the Queensland Department of Harbours and Marine of the incidental capture of cetacea for the years 1962 to 1977 in various places in Queensland. Of the total of 317 dolphins killed, 200 were killed in the last five seasons. (The figures cannot be used to indicate any trend in abundance of animals in the areas as no details of the amount of effort applied in the various seasons was included.) Dr Paterson indicated that the species of dolphin involved was presumably Tursiops truncatus (the bottlenose dolphin) but advised that in September 1977 a baby humpback whale which washed ashore at Surfers Paradise almost certainly died as a result of these mesh netting activities. An adult humpback whale was released from the net there two days before. Dr Paterson expressed doubt about the value of the shark meshing program - although it has a potential value in reducing shark attacks, these have never been frequent when bathers keep to patrolled beaches. In contrast the program has damaged stocks of local marine mammals including dugongs which have frequently been drowned in the nets (Paterson, 1978).

Dr G.E. Heinsohn (1978) referred to assessments of the effects of shark netting programs for the Townsville region. No submissions referred to the extent of the problem in those New South Wales regions where shark meshing is practised.

While at this stage the numbers of dolphins killed cannot be assessed in relation to estimates of total numbers of the species involved, it is unlikely that proposals to halt the netting activities would receive public support, because of general concern about shark attacks. Nevertheless the programs should be examined closely to see if it is practical for any measures to be adopted to reduce the accidental killing of marine mammals.

Summary of Australian stocks

Sperm whales: The most recent assessments for the Division 5 and Division 6 sperm whale stocks suggest that the status of the former stock is far from satisfactory because of overexploitation of the males, but this does not apply to the latter stock. The need for full protection of the Division 5 stock warrants urgent reconsideration of current Australian and International Whaling Commission management provisions. While assessments of the status of the Division 6 stock suggest that it has not been reduced to the same extent as the Division 5 stock, prudence would suggest a much more conservative approach internationally in the determination of catch limits because of the uncertainties associated with stock assessments.

Humpback whales: The evidence from sightings off the west coast suggests it is unlikely that there are fewer humpbacks than in 1963, when assessments suggested that a stock of less than 800 individuals remained of the original

12,000 or more. So far as the east coast is concerned, in 1963 it was estimated that 500 individuals remained of the initial stock of 10,000. The numbers in these stocks remain very small, and there is insufficient information to know whether there has been any recovery.

Right whales: The limited evidence available suggests that there are very few right whales off Australia. Sightings off the west coast are encouraging but do not show a recovery to the extent evident off the coast of South Africa. From reports describing the large numbers of these whales at the beginning of white settlement it can be inferred that recovery to that level, if still possible, will require protection for very many years.

Small cetacea: Most stocks of small cetacea around Australia are not likely to be affected by the limited interference believed to occur. However, assessments should continue of the effect of incidental catches in shark mesh netting and whether preventive measures are practicable.

Whale research

Very few scientists have undertaken research in Australia on cetacea; studies have in general been restricted to those carried out on behalf of the Commonwealth in support of management of commercial whaling activities, or to classification or descriptive work by museums and universities, the latter type of study relying largely on the availability of beached specimens.[1] The size of many cetacea and their inaccessibility make other studies of them difficult and expensive. One line of research receiving special attention at the moment is the assessment of abundance of protected species of large whales by aerial survey. A limited amount of research on the handling of small cetacea in captivity is carried out by private oceanariums.

The Commonwealth Department of Primary Industry has administrative responsibility for fisheries and whaling, but does not itself undertake

---

(1) The submission by the Commonwealth Government Agencies and Departments briefly summarised details of cetacean research in Australia. Other submissions such as those of Bannister, Chittleborough and Heinsohn provided further information.

biological research. It administers the Fisheries Development Trust Account which was established from proceeds of the sale of the assets of the Australian Whaling Commission (see Chapter 2) in 1956. Funds from the account are made available for fisheries research and development, including whale research.

Allocations of $48,000 from 1963 to 1965 and $74,230 from 1971 to 1978 inclusive had been committed from the Fisheries Development Trust Account for whale research (which mainly included biological monitoring at the Albany station, but also contributed to aerial surveys, marking cruises and to a workshop on the determination of ages of sperm whales) (Commonwealth Government, 1978). Between 1972 and 1977 Cheynes Beach contributed $25,000 towards these projects.

CSIRO Division of Fisheries and Oceanography has been the principal government agency carrying out research on whales. From 1951 to 1963 CSIRO monitored the status of Australian humpback whales during the course of Australian land-station operations. From 1963 to 1967 concentration was on sperm whales following the change of whaling activities to that species. Subsequently, with the departure of Mr J. Bannister to the Western Australian Museum, CSIRO contributed annually to the expense of sperm whale monitoring research which Mr Bannister carried out part-time from his new post.[1] Since 1972, when Dr K. Allen joined CSIRO, it has developed a strong program in theoretical population dynamics which has been of considerable importance in the development of current International Whaling Commission models for the assessment of whale stocks.

CSIRO informed us that a full breakdown of expenditure on whale research was not possible but allocations that could be identified specifically totalled $123,200 between 1951 and 1972. This expenditure was related chiefly to the humpback and sperm whale studies but did not reflect the contribution towards work on population dynamics.

---

(1) Until 1970. After this, funds were provided from the Fisheries Development Trust Account.

Since 1976 the Australian National Parks and Wildlife Service has allocated approximately $28,400 for aerial surveys of humpback and southern right whales off the east coast and for a study of the bottlenose dolphin in southern Queensland.[1]

The Office of Regional Administration, Albany, on behalf of the Great Southern Regional Development Committee submitted that 'a research facility at Albany would be capable of undertaking research into all forms of marine life, including whales, and in particular, the potential fisheries known to exist in the Southern Ocean' (Great Southern Regional Development Committee, 1978, p.10).

Their submission suggests that, if whales are allowed to be taken for research purposes, the Cheynes Beach facilities 'would prove an invaluable asset for the proposed research centre'. It also claims that the whale chasers 'could be modified, relatively easily, for use as research vessels'. Similar views were expressed in other submissions to the Inquiry, including those of Project Jonah and Friends of the Earth in Sydney and Melbourne, although these generally emphasised that research should not involve the killing of whales.

The Inquiry is unable to give consideration to the proposal to establish research facilities at Albany, so far as this relates to fisheries, as this is outside its terms of reference. However, it is difficult to believe that the catchers, which are old, steam-driven and designed specifically for whaling, would be suitable for conversion to the modern scientific and oceanographic platform which would be required for effective marine research in the waters around Australia and in the more southern oceans.

It would also seem to be impractical to keep the catchers and processing equipment under maintenance for any future research activities involving the taking of whales, and there would seem to be problems in obtaining

---

(1) The research has been sponsored in support of the Service's general responsibility as adviser to the Commonwealth Government on nature conservation and also to satisfy its statutory responsibility as the scientific authority in relation to the Convention on International Trade in Endangered Species of Wild Fauna and Flora (CITES).

experienced crews for such occasional periods of service. As pointed out by Bannister, if whales were taken from the Division 5 stock to measure changes in pregnancy rates, the numbers involved would be 'in the low hundreds...at regular intervals (of), say, three to five years' (Inquiry Hearings, p.497). Leaving aside concerns about the state of the stock, numbers of this order would not justify maintaining a whaling capacity.

While Albany is conveniently placed in relation to sperm whale migration paths, it is unlikely to be an ideal place for equally pressing studies such as social behaviour of whales in breeding areas, where long-term observations of less transient groups would be desirable. It has been suggested that this would be best done from places (perhaps oceanic islands) where the continental shelf is much closer to the coast.

One aspect of the Albany operation that would however be valuable for future research is the expertise in aerial survey and identification of whales that the spotter pilots and gunners have developed in the course of their long association with whale chasing. Their participation in future surveys would make results more comparable and avoid the need for familiarisation with the techniques.

A line of cetacean research that may require support is the study of stranded specimens. This depends on effective controls to protect stranded specimens from vandalism and a system for prompt notification of people able to preserve specimens. The Inquiry considers that legislation providing for the protection of dead as well as live cetacea (as already exists in most States), and periodic public relations campaigns to inform the public of the importance of such specimens, would be a useful aid to the museums and other institutions pursuing such studies.

Despite the significance attributed by many submissions - including by academics - to the whale's brain and intelligence, the Inquiry was not informed of any private or government research in Australia on matters such as neuroanatomy, the level of sophistication of communication between cetacea, or their behaviour. The Inquiry supports in general the attitude expressed by the Australian Conservation Foundation in its submission that more attention should be given to the study of live whales, a field where improved techniques are rapidly developing.

There are so many gaps in man's knowledge of whales that further biological research would almost certainly assist in measures for their conservation. However, only in special circumstances could we support taking live whales for this purpose; generally the research could be carried out on stranded specimens, and it may also be possible to take samples from live whales without harming them. It has been suggested that biological studies of whales may be valuable in research such as that on cancer, the human metabolism, and the effect on animals of ultrasonic beams (Project Jonah, 1978, pp.3B/94-7).

The need for and nature of future research in Australia on cetacea was discussed at the Melbourne public hearings on 23 August 1978. In addition various submissions, such as those of Project Jonah, the Australian Conservation Foundation, Dr M. Bryden, and Dr E. Guiler, included references to the inadequacy of current research programs in Australia.

Nevertheless, in the present financial climate the Inquiry hesitates to recommend an expansion of government funding for general cetacean research. The current level of funding should however be maintained, despite the fact that research in support of continued local exploitation is no longer necessary. The strong popular and scientific interest in and concern for whales would in our opinion certainly justify this. We would encourage interested organisations and community groups to help in providing additional support for programs of research on cetacea.

Some areas of research will be more important than others. We consider that the following matters should be given priority.
    (a) Aerial surveys should be continued and conducted so as to permit comparison with previous commercial spotting operations. These should be undertaken at appropriate intervals so that trends in the abundance of sperm, humpback and right whales can be monitored. Migration patterns and social behaviour should also be studied. This research would enhance the general understanding of whale populations and their dynamics which, whether or not whaling

continues, is required for their conservation. It would also assist in examining issues such as the possible need to give special attention to breeding or nursery areas.

(b) Development and refinement of population models and techniques for assessment should continue and be extended to take account of the total ecological relationships of whales.
While whaling continues, substantial progress should be made in overcoming some of the criticisms of current procedures. If whaling ends internationally in the future, work on the modelling and assessment of whale populations would need to continue both because of the rapidly expanding commercial interest in the resources of the Antarctic and the role of whales in other marine ecosystems.

(c) Australian strandings of cetacea should be monitored. If more of these specimens can be made available intact to scientists, valuable biological information can be obtained. This should avoid much of the problem of taking live specimens for scientific purposes.

Australia's contribution to international research has been particularly significant in the field of population modelling in recent years. The Inquiry, taking the view that Australia should remain a member of the International Whaling Commission whatever attitude it adopts towards whaling (see Chapter 5) considers it most important that Australia's experts in this field continue to participate actively in the Commission to improve the reliability of scientific advice, and to ensure that Australia's interests are well represented.

7. WHALE PRODUCTS AND ALTERNATIVES

The great whales have always been hunted for the products they provide rather than because they compete with man for the resources of the sea. Today baleen whales are hunted for their meat and edible oils; sperm whales for their oil which is used mainly for lubrication and tanning leather. A detailed summary of products obtained from whales, their chemistry and their applications is contained in Appendix 13.

The Inquiry received many submissions claiming that all products obtained from whales could be replaced by available substitutes. We did not attempt to examine this claim in detail for all parts of the world. However, noting that the United States (in 1971) and New Zealand (in 1975) have banned imports of all whale products, and that in 1973 the United Kingdom banned imports of all products except sperm oil, spermaceti and ambergris, it is reasonable to assume that substitutes are widely available. There may of course be economic or technological reasons why some countries could not switch quickly to a general use of substitutes, and there are also some communities which would wish to continue to take whales for social and cultural reasons. We take these points into account when making recommendations on the general attitude which Australia might adopt towards whaling.

The more directly important questions for the Inquiry are whether there are substitutes available for whale products used in Australia and whether there are any difficulties in adopting such substitutes. Related to this is the proposal advocated in many submissions that Australia should ban the import of all whale products so as to hasten the end of commercial whaling by other countries.

These are matters of some substance, particularly now that Cheynes Beach has confirmed that it will not be whaling after the 1978 season ends, and the Inquiry therefore conducted an extensive independent investigation on products used and the availability of alternatives in Australia.

Sperm oil, spermaceti, whale meal, whale solubles, ambergris and ivory are all whale products used in Australia although 'most of the argument in relation to substitutes has revolved around the processed end products from crude sperm oil' (Cheynes Beach, 1978, p.34). A summary of the products and quantities produced by Cheynes Beach for the last two years is listed below.

Table 7.1
Whale products produced by Cheynes Beach

	1976	1977	
Sperm oil	5741	3931	tonnes
Whale meal	1372	1060	tonnes
Whale solubles	2508	1754	tonnes
Teeth (ivory)	3849	2300	kg
Ambergris	381	211	kg

Source: Annual Whaling Inspector's Reports, 1976 and 1977 Sperm Whaling Season.

A very high proportion of Cheynes Beach's oil production is exported as commercial crude sperm oil which is a mixture of sperm oil and spermaceti. Over the period 1975 to 1977 overseas sales represented 91 per cent of the total - see Table 8.1. Most of the sperm oil sold in Australia has been filtered to remove the spermaceti.

All spermaceti used in Australia is imported from overseas, the quantities involved (28 tonnes in 1977-78) being too small to justify local refining. Small quantities of sperm oil are also imported (5 tonnes in 1977).

The use in Australia of whale meal and solubles, ambergris and ivory is linked to Cheynes Beach's production. As explained in Appendix 13, no difficulties would arise if these products were no longer available. Nor would there by any problems if imports of baleen whale products - which are used very little in Australia anyway - were banned. The focus must thus be on the applications in Australia of sperm oil and spermaceti.

Sperm oil

The chemical composition of sperm oil classifies it as a liquid wax that makes it unique among readily available commercial fatty oils.[1] Sperm oil and products derived from sperm oil have been used in many applications but only two of these are of major importance; lubrication and leather tanning. Over 85 per cent of sperm oil sold by Cheynes Beach in Australia was used in the lubrication and leather tanning industries in 1977.[2]

The search for substitutes for sperm oil started in earnest in the United States after the United States Endangered Species Conservation Act of 1969 went into effect in June 1970. At that time eight species of whales including the sperm whale were placed on the US list of endangered fish and wildlife by the Department of the Interior, and on 2 December 1970, a ban was placed on imports of oil, meat and other products from these species of whales. Special government import permits allowed quantities of sperm oil to be imported as an interim measure until 2 December 1971 when interstate trade in whale products was completely banned in the United States (Recchuite, 1973; US Congress HR 3465, 1975). This ban naturally led to intensive research for sperm oil replacement products. Other factors over the last decade have been declining world supplies of sperm oil, the possibility of further market restrictions in Britain following bans on imports of all whale products except sperm oil, spermaceti and ambergris, and the New Zealand import ban.

Lubrication: Sperm oil has several characteristics that make it an ideal lubricant or lubricant additive. It is extremely 'slippery' giving

---

(1) Liquid waxes do not occur commonly in animals and plants. Liquid waxes are found in the sperm whale, the bottlenosed whale, some species of dolphins, the mutton bird, one petrel species, the seed of the jojoba plant and several fish and squid species (Warth, 1947).

(2) Information provided by Australian buyers of sperm oil revealed that in 1977, 205 tonnes were used for lubrication and 99 tonnes were used in leather tanning. Australian sales totalled 351 tonnes and 5 tonnes were imported.

excellent antiwear protection to metal surfaces. Sperm oil can be chemically modified to improve its antiwear properties greatly under extremes of temperature and pressure. Sulphurised sperm oil is the most important chemical derivative used in the lubrication industry. It is more soluble in paraffinic (mineral) oils and more chemically stable than most other sulphurised animal and vegetable oils. The submission from Project Jonah summarised the position by saying:

> Since H.G. Smith's discovery in 1939 (that sulphurised sperm oil was more soluble and stable than sulphurised lard in paraffinic oils for extreme pressure (EP) lubrication) sperm whale oil has been used primarily in lubrication applications. Over 5 per cent of use in the United States prior to the ban on sperm oil was in lubrication. Sperm oil's metal working properties, oiliness, and non-drying characteristics allow it to be used directly in lubrication; however, it is more important as a chemical intermediate as it can be sulphonated, sulphurised, oxidised, sulphur-chlorinated, and chlorinated to produce products for use as wetting agents and extreme pressure (EP) additives. The sperm oil mainly acts as a carrier for sulphur or the other intermediates which provide the EP lubrication qualities. The primary uses of these EP lubricants have been as automatic transmission fluids, metal working oils, and industrial and automotive gear oils and greases (Project Jonah, 1978, p.3E/7).

The oil from the jojoba shrub (Simmondsia chinensis) would be a suitable replacement for sperm oil in almost all applications because of its similar chemical composition and many of the submissions received by the Inquiry listed jojoba oil as the available substitute for sperm oil. Unfortunately the world availability of jojoba is very limited and is unlikely to increase significantly in the next ten years. We note that interest has been shown in growing jojoba in Australia, with the sale of seeds and seedlings being actively promoted. Appendix 14 provides more detail on the chemical and commercial feasibility of using jojoba oil as a substitute for sperm oil.

Replacements for sperm oil in lubricants, developed by American companies, are generally available in Australia through subsidiary companies and agencies. In addition in recent years a number of Australian chemical companies have formulated proprietary products as replacement additives for sulphurised sperm oil in lubricant applications. Many of these have been accepted and others are currently being tested by Australian oil companies.

As pointed out in the submission by Dr I. Furzer, Senior Lecturer in Chemical Engineering at the University of Sydney, 'it would be possible to synthesise all the individual components of sperm oil using available petrochemical technology, and by blending produce a synthetic sperm oil. The cost of this synthetic oil would be high due to the large number of components that must be produced' (Furzer, 1978, p.2).

The approach adopted instead has been to select and blend many different chemical substitutes to obtain the required lubrication properties for particular uses. Thus, for a single sulphurised sperm oil product a large number of proprietary products have been developed from various esters[1] of animal and vegetable oils, fatty chemicals and petrochemical compounds (Recchuite, 1973).

These needed to have the following properties:
(a) resistance to chemical oxidation;
(b) thermal stability;
(c) solubility in mineral oils used as the base for lubricating oils and greases; and
(d) excellent metal lubricating characteristics even at high temperatures and pressures.
Of course cost comparability was also an important consideration.

For some uses of lubricants, 'sperm oil' became a sales tool used to promote certain proprietary products and so sulphurised sperm oil found its way into many oil formulations where its special qualities were not essential. As research for a sperm oil substitute continued, sulphurised lard oil and vegetable oils were put back into these formulations (Recchuite, 1973). For most metal-cutting and metal-drawing tasks in Australia, locally manufactured triglyceride oils and synthetic fatty

---
(1) An ester is the general name for the chemical compounds produced when an organic acid reacts with an alcohol.

esters, either separately or in combination, have replaced sperm oil at prices ranging between $450 and $500 per tonne. These prices are significantly lower than those for the equivalent sperm oil products, which compensates for a lower resistance to oxidation.

In 1972 the General Services Administration in the United States determined that sperm oil was no longer a strategic material because there were substitutes. The British Ministry of Defence, after reviewing the use of sperm oil in defence products, decided in March 1978 that lubricants containing sperm oil derivatives used by the Royal Small Arms Factory Enfield, Royal Artillery units and other sections of the defence forces would be replaced by the end of 1978.

Dr J. Gilbert, the British Minister of State, referring to the use of sperm oil derivatives in the British armed services in a statement to the House of Commons on 6 July 1978 said:
> In all cases the exclusion of sperm whale oil is being achieved at no cost penalty to the Defence Budget nor will the operational effectiveness of the services be in any way impaired.

The Australian Minister assisting the Minister for Defence has said that sperm oil or sulphurised sperm oil is not a specified requirement for the Australian Services.

Dr I. Furzer, referring to his correspondence with the nine major oil companies within Australia, summarised their position as follows:
> Some of these oil companies had phased out sulphurised sperm oil as early as 1974. All these companies, with their headquarters being overseas with a few exceptions, have had access to US technology because these synthetic materials are used in America. The main situation of all these nine companies is that they are all very conscious of a possible ban on sperm oil and have taken steps to find substitutes. Many companies have phased out in the intermediate period and up to mid-1978...there were few companies using sperm oil (Furzer, Inquiry Hearings, p.247).

The Australian Institute of Petroleum presented a similar picture in a letter to the Inquiry on 21 June 1978. It wrote:

> It should be noted that some marketers of lubricants in Australia do not include sperm whale oil or its derivatives in any of their products having phased out the use of these materials in recent years.
>
> Among the companies that still do, there does not appear to be complete uniformity in the range of their products which contain sperm oil or its derivatives.
>
> However, most of these companies have informed us that they are phasing out the use of sperm whale oil and/or are carrying out laboratory work aimed at finding, or proving the suitability of, synthetic substitutes.
>
> We should add that, in the opinion of one company, the development of satisfactory substitutes, for inclusion in at least some of the lubricants concerned, appears to pose considerable technical difficulty, quite apart from any economic considerations.

It is evident that substitutes exist for sperm oil and sperm oil derivatives in all major uses of lubricants. The situation is less clear for a small number of particularly demanding uses where some manufacturers still consider sperm oil derivatives essential to obtain the required performance. Information obtained from suppliers of lubricants and sperm oil substitutes strongly suggests, however, that it is possible to replace sperm oil derivatives in all cases although this will involve some increase in cost and may take up to two years to complete the required development and testing of new formulations.

The United States experience supports this conclusion. All imports of sperm oil have been banned there since December 1971 and, while there were considerable stocks of oil in 1971 that have been available to American industry, sperm oil has now been almost completely replaced by substitutes. All current exemptions to the provisions of the Endangered Species Act 1973 will expire by mid-1980, making the possession, sale and transporting of sperm oil illegal (US Public Law 94-359).

The fact that a number of oil companies have already replaced sperm oil derivatives over a wide range of products suggests the cost of changing over is not substantial. However, it is generally recognised that the substitutes for sperm oil derivatives that must be used where the highest standards of performance are required are currently more expensive. As an example, in one application noted by the Inquiry the substitute for sulphurised sperm oil is double the price, but because sperm oil is used only as an additive, substitution will result in an estimated increase in the total cost of the product of only 15 per cent or 7¢ per kilogram.

The replacement of sperm oil will also require time for some lubrication specifications to be amended. In its letter to the Inquiry of 21 June 1978 the Australian Institute of Petroleum pointed out that:

> Some manufacturers of vehicles (and possibly industrial machinery) still specify, for particular functions, the use of lubricants containing sperm whale oil or derivatives thereof...It also follows that substitution does not depend only on the development of alternatives which lubricant marketers regard as satisfactory but also, at least in some instances, on the acceptance of such alternatives by vehicle (or machinery) manufacturers and their amendment of lubricant specifications accordingly.

Obviously where existing lubricant formulations perform to manufacturers' specifications, some users of lubricants are reluctant to test or consider alternatives without marketing pressure to do so. The testing of alternatives involves time and expense because of the range of substitute proprietary products available and the variable performance of additives in different base oils and greases. While complete replacement of sperm oil in Australia will take some time and will involve some increase in costs, particularly for those oils and greases prepared for specialised applications, from the information it has obtained the Inquiry concludes that the technology exists for the replacement of sperm oil and its derivatives in all cases where it is used in lubrication.

<u>Leather tanning</u>: Leather is an animal hide or skin that has been processed for use by man. Hides or skins in their natural state taken from animals are composed of proteins and water which decay quickly unless preserved.

The manufacture of leather from raw hide first requires the removal of hair or wool and mucoproteins leaving a concentrated network of protein fibres interspaced with water. These are chemically stabilised by the action of tanning agents, principally chromium salts, that displace the water and then combine with and coat the protein fibres. The stabilised leather is then 'fat liquored' by churning the wet leather with oil emulsions to prevent the leather becoming hard when it dries.

The importance of using an oil with the correct properties is explained by the British Leather Manufacturers Research Association in their report to the British Department of Industry in 1976:

> The choice of lubricant is critical in determining the tactile characteristics of leather, such as drape, handle or softness, and aesthetic properties such as surface sheen combined with freedom from greasiness...In order to develop the required characteristics in a given leather, it is common practice to use blends of different oils, which may be formulated within the tannery or purchased as proprietary products from chemical companies...
>
> ...Lubricants should be resistant to physical or chemical change during the service life of consumer products made from leather. For example, they should not change colour, or harden, on exposure to air and light, they should be relatively odourless and should not migrate within the leather to give surface efflorescences or 'spues'. Furthermore, they should not interfere with the fabrication of articles made from leather, e.g. by impairing the strength of adhesive joints or promoting hydrolytic degradation of the leather during wear or storage under humid conditions (BLMRA, 1976, p.1).

Sperm oil meets most of these criteria and it has proved very difficult to find a synthetic blend to replace it that has the same wide range of performance characteristics required for fat liquoring. The lack of any comprehensive theory of the physics and chemistry of fat liquoring has meant that replacements for sperm oil have to be developed by trial and error.

In its 1976 report the British Leather Manufacturers Research Association concluded:

Overall, we conclude that the leather industry would be faced with considerable technical problems if fat liquors based on sperm oil were to be withdrawn from the market...

There remain substantial areas of leather manufacture where substitutes are unable, on the basis of present knowledge, to give process and/or product performance comparable to sperm oil based products (BLMRA, 1976, p.37).

The areas of manufacture referred to are mainly the production of high quality clothing leathers for which the British leather industry is noted. The report listed:
(a) the production of high quality gloving leathers;
(b) the production of special surface effects on suedes and full grain leathers;
(c) the development of softness without greasiness in coarse hides such as bovine and sheep skins.

More recently, in a letter to the Inquiry of 27 June 1978, Mr R.L. Sykes, a Director of the British Leather Manufacturing Research Association, said:

It is, I think, fair to state that the more recent attempts by some speciality oil manufacturers are a considerable advance on the products they were offering three to four years ago in so far as leather peformance is concerned, although certain tanners have indicated that there are difficulties in ensuring consistent application...

...(However) It is my impression that whereas the quantity of sperm oil used could be reduced by the blending of sperm oil with other oils, it would not be possible at this time to produce the full range of leathers manufactured in the UK to current expected performance levels if sperm oil were not available to the leather industry.

Turning now to the situation in Australia, in a letter to the Inquiry on 23 June 1978 the President of the Federated Tanners' Association of Australia, Mr E.G. Francis, said that while sperm oil was used in fat liquors in Australia, particularly in the manufacture of leather for handbags, garments and footwear uppers, 'sperm oil substitutes are now being used extensively and in my Company's plants we no longer use sperm oil and successfully use substitute synthetics'. In summing up Mr Francis said 'I can state quite categorically that there are no problems in using the substitutes except for a minor increase in cost'.

Not all experts agree with this view. Dr T.G. Scroggie, leader of the Leather Research Group in the Division of Protein Chemistry, CSIRO, summarised some difficulties in a letter to the Inquiry on 2 August 1978:

> I believe that opinion would differ between different tanneries in the industry on the necessity for use of (sperm) oil and there is insufficient basic information available to allow a conclusive answer to this question. The situation is complicated by the fact that, without running a series of analyses on the oils used by the industry, we cannot assess accurately the extent of use of sperm whale (or any other) oil at present. This is mainly because many of the oils are purchased under trade names which do not disclose the nature of the oil and the supply companies are often reluctant to give this information either to the tannery or to a research organisation such as ours.

The Inquiry is not qualified to assess independently the wide range of products offered as alternatives to sperm oil and its derivatives as fat liquors and determine the exact consequences for the Australian leather industry of a removal of sperm oil from all fat liquor blends.

However, based on the information available to it the Inquiry considers it will be possible to replace sperm oil completely in the leather industry although it may involve some increases in fat liquoring costs and adjustment difficulties in producing high grade leather products. We consider that two years would be sufficient to carry out the required development and testing of new fat liquor formulations.

<u>Other applications</u>: Sperm oil and its derivatives are or have been used in a variety of other applications in the chemical, pharmaceutical, cosmetic and textile industries but no claim has been made that sperm oil is essential in any of these.

Crude sperm oil is sold throughout Australia for use in the preparation of fish bait lures. Its use for this purpose is not essential and alternatives such as fish and lard oil mixtures can be successfully used.

It was claimed that sperm oil was necessary for the culture of some penicillin strains (Smith, 1978) but the sole manufacturer of penicillin in Australia, the Commonwealth Serum Laboratories, stated in a letter to the Inquiry on 28 June 1978 that no sperm oil was used for the commercial production of penicillins in Australia.

Spermaceti wax

Spermaceti wax is imported from the United Kingdom and Europe. Consumption in the 1977-78 financial year totalled 28 tonnes.

Table 7.2
Australian consumption of spermaceti wax

Year	Quantity (kg)	Value ($)	Unit value (c/kg)
1973-74	7 697	3 000	39
1974-75	10 464	9 000	86
1975-76	15 173	14 000	92
1976-77	36 131	34 000	94
1977-78 (preliminary)	28 394	35 578	125

Source: ABS Australian Trade Statistics

Spermaceti is an ideal ingredient in cosmetics because it resists oxidation and does not develop odours in emulsions or in its original form. It provides a non-greasy, non-toxic base for a wide range of cosmetic products and, in particular formulations, spermaceti can be used to increase the opacity and viscosity, raise the melting point, or soften the texture of the final product.

However, alternative formulations which do not contain spermaceti are available for all of these cosmetic products (Jellinek, 1970; Sagarin,

1957) and a number of companies in Australia produce a wide range of cosmetics without using whale products (Project Jonah, 1978).

Spermaceti can also be used as the wax component in a wide range of industrial products such as polishes, candles, matches, crayons and chalks, batteries and food coatings, but a number of other waxes could be substituted

There are a number of products synthesised from vegetable and animal oils and petrochemicals that are currently being sold within Australia as replacements for natural spermaceti.

Trade prices for synthetic spermaceti vary between $2.00 per kg and $5.00 per kg compared with a range of between $2.10 and $3.40 per kg for natural spermaceti. Part of this variation in the price for synthetics can be explained by differences in chemical formulation and melting point, but it appears that synthetic spermaceti can be obtained at prices comparable with natural spermaceti. Products such as cetyl palmitate and myristyl myristate, which are available in Australia, can also be used as alternatives to spermaceti in some cosmetic and industrial uses.

The Inquiry considers that there are suitable alternatives to spermaceti in all its applications but recognises that should natural spermaceti be no longer available, changes in present cosmetic and industrial formulations will require time for testing and approval.

Conclusion

The only whale products used in Australia which may not be readily replaceable are sperm oil and spermaceti, and even for these there are substitutes available for most uses to which they are put. For those cases for which sperm oil is still required the Inquiry considers that a period of two years would suffice to develop and test adequate substitutes - particularly since developmental work is generally already well in hand. The cost of using substitutes would not be an important factor for most applications.

We also cannot accept the argument raised against the use of substitutes, that they involve a formulation of mineral oils derived from fossil fuels, a non-renewable resource. Substitutes can be obtained from vegetable oils, and in any event the total supply of sperm oil upon the Australian market is so small in relation to the Australian oil consumption of 600,000 barrels a day that the question of use of non-renewable energy cannot really be considered relevant (Furzer, Inquiry Hearings, p.262).

It would therefore be possible for Australia, if it wished, to indicate that after a two-year adjustment period the import of sperm oil, spermaceti and all other whale products would be prohibited. This would be a similar approach to that used in the United States, where imports of sperm oil were allowed under licence for one year as an exception to a general ban on imports of whale products. It would however seem unnecessary and undesirable to restrict imports into Australia during the adjustment period. Only a small amount of stockpiling would be expected because of the costs involved, while the option to build up stocks would help to avoid any difficulties which might arise in very specialised applications.

8. THE CLOSURE OF CHEYNES BEACH

When the Inquiry began, 'whether Australian whaling should continue or cease' was the central issue in its terms of reference. One possible outcome was a recommendation that Cheynes Beach's whaling licence should not be renewed. This would have forced the whaling station to close. The announcement by the company that whaling operations would stop in the near future, since confirmed as the end of the 1978 season, resolved this question and so changed the direction and emphasis of the Inquiry.

Had the Inquiry recommended the closure of Cheynes Beach's whaling operation, the question of compensation would have required consideration. It was a feature of many of the submissions opposed to the continuation of whaling that they argued forcibly for adequate compensation for those individuals directly affected by the closure of the whaling station. Cheynes Beach itself said in its submission to the Inquiry that: 'should the company be forced to stop whaling it would expect compensation to be paid both to the company and its employees'. The basis on which this claim was raised is no longer applicable.

However, some people, including particularly Mr Barr and Mr Wells acting as spokesmen for the unions to which most of the employees belonged and Dr Mosley representing the Australian Conservation Foundation still argued that some compensation should be paid (Mosley, Inquiry Hearings, p.820; Barr, Inquiry Hearings, p.134; Wells, Inquiry Hearings, p.123). The Inquiry therefore feels obliged to make some comments on the matter.

The demand for sperm oil

The statement to shareholders announcing the intended closure gave the 'sharp down turn in the demand for crude sperm oil' as the basic reason for the decision (Cheynes Beach Press Release, 31 July 1978 - see Appendix 4) To state further the substance of the press release, the directors believed operations this year (1978) would result in a substantial loss, and it was unlikely there would be any profit in whaling in 1979. There was a serious move from filtered sperm oil to alternatives. The reason given by the Chairman of Directors was the doubt of continuity of supply from Australia, being an unforeseen effect of the Inquiry. Hitherto, he said, Australia had fulfilled a substantial proportion of the free world demand. From the early part of this year, the company's buyers had shown extreme reluctance to commit themselves to forward buying. At the time of the press release only 1000 tonnes had been sold out of the year's operations, but sales were slow and prices were below the cost of production.

The Inquiry sought further information upon price trends from Highgate and Job Ltd, the major processor and refiner of sperm oil in the United Kingdom. The correspondence is set out in Appendix 15.

First we shall refer to the sales over recent years of Australian sperm oil. Sales of sperm oil accounted for nearly 80 per cent of Cheynes Beach's revenue from whale products (Cheynes Beach Press Release, 31 July 1978) and most of the sperm oil produced by the company was sold to overseas buyers. Table 8.1 gives the Australian and overseas sales of sperm oil in recent years. The world demand and world price of sperm oil have therefore had an overriding effect on the company's income and profitability.

Table 8.1
Sales of Australian sperm oil production

Year	Sold within Australia (tonnes)	Sold to overseas buyers (tonnes)
1975	396	2762
1976	495	7137
1977	351	2206
1978 (provisional)	350	4500

Source: Harrisons and Crosfield (Cheynes Beach's agents for the sale of sperm oil)

In looking at the international market for sperm oil, the following extracts from the letter of Highgate and Job Ltd are particularly pertinent:
During the (ten years preceding 1970), the price variation (for sperm oil) was from £45 per tonne up to £120 and back to £90. This covered

a normal sequence with the £120 per tonne price being, at that time, considered a very excessive peak, at which stage tonnage began to be lost by the users showing resistance.

However, with the introduction of the seventies and the inflationary period that followed, a new sphere of prices naturally ensued.

Between 1970 and 1975 the price modulated between £90 per tonne and £175 per tonne, which was a fairly constant figure when placed against supply and demand and this structure was favourably accepted by the end users of sperm oil. It was acceptable at that time as it showed an advantage over the synthetic prices ruling on that day.

Between 1976 and 1977 an exceptional upsurge occurred in the price of crude sperm oil ranging from the £170 operating in 1975 to £320 in 1976 and £470 in 1977. This was an unfortunate sharp increase and it was due to several factors -

(a) The International Whaling Commission in June 1976 severely reduced the quotas for the southern hemisphere resulting in a panic position being created.
(b) Due to this reduced tonnage South Africa decided to cease production.
(c) At this time the cost of mineral oil rose sharply, thus increasing the costs of production to the whalers.

All these factors tended towards this sharp increase which in the end proved to be the undoing of the sperm oil market as a very strong consumer resistance was set up.

In 1977 and 1978 a reactionary movement took place due to end users' resistance and the price has now fallen to about £250 per tonne.

At this revised level consumers are very interested in the use of this oil but due to the uncertainty of the situation over the preceding years, a favourable market no longer exists as the raw material has lost interest with those who depended upon it most.

Due to this uncertainty a trend has been created to investigate the use of synthetics even though it is recognised that the synthetic material cannot produce a finished product equal to that from sperm oil...

It is anticipated that sperm oil will still be available from Japan, Iceland, Peru and perhaps Chile so that the international market will still be served from these sources (when Australian whaling ends) (Highgate and Job Ltd, 1978 - see Appendix 15).

A summary of quarterly average prices for crude sperm oil between 1970 and 1978 is given in Table 8.2.

Table 8.2
Crude sperm oil prices
(£ sterling per tonne)

	Quarter 1	2	3	4
1970	95	125	125	140
1971	155	150	140	115
1972	115	105	115	125
1973	120	140	155	150
1974	180	205	185	185
1975	180	175	175	150
1976	230	320(a)	340	370
1977	390	420	470	400(b)
1978	275	250		

(a) Quota was cut, sperm oil supplies were reduced and prices increased.
(b) The stage at which end-users (buyers) resistance became most evident.

Source: Highgate and Job Ltd

Cheynes Beach provided some comments on the letter of Highgate and Job Ltd; these are set out in full in Appendix 15. The main points made by Cheynes Beach are that in February 1978 it was widely anticipated that, as had happened many times before, the demand for crude oil would once again show some improvement by mid-1978, and provided the uncertainty of supply could be contained, the market would continue. To overcome this difficulty, it was added, the company was in the process of negotiating direct with a major processor of crude oil for a joint venture to ensure stability of the industry, and as part of the proposal the processor was to spend A$300,000 to provide modern processing and refining equipment. (This was not referred to in the press release announcing Cheynes Beach's decision to end whaling.) However it was the company's view that it was further uncertainty as to future supplies, resulting as an unforeseen effect from the appointment in early 1978 of the Inquiry, which was 'the straw which broke the camel's back'.

Our strong impression left by the sequence of events is that the fall of the market was much more than a normal down turn, and Cheynes Beach certainly had a task in front of it to reverse that trend. In our view, this is borne out by the fact that it was considered necessary for the company to announce the closure to enable a sale to be negotiated for the rest of the year's production, and that the sale was eventually negotiated at a greatly reduced price of US$510 (about £260) (Cheynes Beach - see Appendix 15).

The Inquiry has given consideration to the developments in the sperm oil market and the views expressed by both Highgate and Job Ltd and Cheynes Beach. The main factors which have been suggested as influencing these developments are:
- (a) recent high prices for sperm oil products and a history of uncertainty within the sperm oil market;
- (b) continued downward revisions of sperm whale quotas by the International Whaling Commission since 1973;
- (c) continued pressure by conservation lobby groups within Britain to ban imports of all whale products into Britain;
- (d) the establishment of the Inquiry into Whales and Whaling within Australia, one possible outcome of which was the ending of Australian whaling which would affect future supplies to Britain;
- (e) concern expressed during the International Whaling Commission meeting in June 1978 by Australian scientists at a fall in the pregnancy rate in the Division 5 stock and, therefore, as to the state of the stock.

The situation facing Cheynes Beach in 1978, as the Inquiry sees the situation, was that a substantial increase in world prices for sperm oil was required to make whaling in Australia profitable again, but any such increase in prices would make substitutes for sperm oil even more attractive. Any further shift to substitutes in the face of higher prices would reduce world demand and tend to depress prices further.

While it is obviously impossible to determine the exact extent to which individual factors influenced the demand for sperm oil produced by Cheynes Beach, the Inquiry considers that the market trends leading to the decline in prices were firmly established by the latter half of 1977. With quotas declining and operating costs rising, particularly labour and bunker fuel costs, it does seem that the crisis that Cheynes Beach faced in 1978 would have arisen in any event.

An early end to whaling in Australia seems to have become a certainty following further evaluation of whale sighting data for the Division 5 sperm whale stock. Australian scientists have now reached the conclusion that the stock should be protected and that it would be highly desirable that the 1979 catch limit for it be reduced to zero. This catch limit will be reviewed at a special meeting of the International Whaling Commission and its Scientific Committee during November and December 1978.

Compensation

Compensation for closure can only be considered in respect of a business or employment which would otherwise have continued except for some government action. The Inquiry has a natural concern for the company's employees, many of whom have had long periods of service with the company. But, for the reasons we have set out, the industry would inevitably have come to an end after the current season, and the Inquiry therefore feels obliged to record that it sees no basis upon which it can recommend that compensation be paid either to Cheynes Beach or to its employees.

The closure of Cheynes Beach might be compared with situations that have arisen in the past in the United States and Canada. In the United States, although the Endangered Species Conservation Act of 1969 did not provide for domestic whaling controls, to be consistent it was decided not to renew a whaling permit at Richmond, near San Francisco. This station had already been reduced to one catcher operating almost solely on scientific permits and was hardly economic. A claim for compensation failed (US delegation to International Whaling Commission Meeting, 1978).

A ban on commercial whaling in Canada was announced by the Minister of Fisheries in 1972. Whaling on Canada's Pacific coast had already stopped in 1967 because of economic factors but was continuing on its Atlantic coast from three shore-based whaling stations in Newfoundland and Nova Scotia. The International Whaling Commission quotas for the 1973 season were set to a new low of 143 fin and 70 sei whales. Concern about the falling numbers of these whale stocks and doubt about the future economics of operating with such small quotas led to the ban. At the time of the closure the Minister explicitly stated that the Crown accepted no liability to pay compensation but decided to make ex gratia payments, because for some stations the ban on whaling would have affected the continued existence of rather isolated fishing communities, which were entirely dependent on the fishing and whaling industries (Canadian Government, letter to the Inquiry, 28 August 1978).

In the Australian case, to make ex gratia payments to Cheynes Beach and its employees would be to single out the whaling industry for preferential treatment, when a number of Australian companies have had to close down because of changed economic conditions in recent years.

9. WHALE BRAINS, ANATOMY AND BEHAVIOUR

The large brain of the whale and the striking features it shares with man's brain have fascinated scientists for many years. So too has the behaviour of whales in the open seas and of dolphins in captivity.

The Federation of American Scientists, an organisation representing more than 700 scientists, has stated:
> In the twentieth century man has generally accepted his evolutionary descent from lower animals. But he has not yet accepted an obvious corollary to Darwinian evolution - a spectrum of emotional awareness and of intellectual ability among the animals. This blindness must eventually fail (FAS Public Interest Report, October 1977, in Project Jonah, 1978, pp.3B/21-2).

Another interesting statement is that of Alfred A. Berzin, Chief of the Cetacean Research Laboratory, TINRO, USSR:
> (The sperm whale) is undoubtedly an animal with a cortex of complex structure corresponding to complex psychic manifestations. The sperm whale brain must possess an extreme functional plasticity and practically inexhaustible possibilities for establishing links between stimuli and the form of reactions. The sperm whale brain structure is such that this can be said to be a 'thinking' animal capable of displaying high 'intellectual abilities' (Berzin, 1972, in Project Jonah, 1978, p.3B/20).

The International Whaling Commission has recognised that the behaviour and intelligence of whales are considerations that cannot be overlooked, and is actively seeking 'information on behavioural studies in relation to assessment and management' (Chairman's Report, 1978 IWC meeting).

Many submissions to the Inquiry in favour of conservation of whales were based on their 'intelligence'. Historically whales have been seen as intelligent. This chapter discusses the scientific basis of the assertion of intelligence, setting aside for later consideration its relevance to the issue of whaling.

What do we mean by intelligence when we speak of whales? It is not a new development for the concepts of reasoning and thinking to be introduced into animal behavioural studies. It is clearly the basis of the experiments teaching chimpanzees and other apes to communicate using American Sign Language for the Deaf. Certainly intelligence comprehends more than mere adaptation to the environment, a quality which seems to be accounted for sufficiently in the animals' capacity for survival. Dr Myron Jacobs, a cetacean neuro-anatomist and Associate Professor, Department of Pathology, New York University College of Dentistry, wishes to avoid the anthropomorphic quality of the term 'intelligence' which he says should really be restricted to man (Jacobs, 1972). So he suggests the level of cortical activity of the brain as a test.

This leads to the question of what use the whale makes of its brain capacity. Any study of the whale's intelligence is obviously made difficult not only because its marine environment is so different to man's, but also because of the whale's limitation in having 'no hands to grasp or manipulate objects', and because body movement and sound generation remain the only two external motor responses of the cetacean brain (Morgane, 1974; Jacobs, 1974). Another aspect of this problem is that the sea, with, for example, its buoyancy and relatively stable temperatures, does not seem a demanding enough habitat to require an advanced brain (Bunnell, 1974).

Professor Jerison, a neuro-psychiatrist and Professor of Psychiatry of the University of California, at Los Angeles, has a different approach to the concept of intelligence. He does not accept the view that the different environments of different species make a common definition of intelligence impossible. In the known 150-million-year histories of birds and mammals there has been an enlargement of the brain which, to Jerison, suggested that it was required as an unusual adaptation. It seemed to him that the idea of intelligence had to be introduced, and that the adaptation was a solution to the problem of handling an

overwhelming load of information from the senses and feedback systems of
the brain. Putting it very simply, the mechanism to cope with this
information was provided by enlargement of the pathways of the brain.
Jerison therefore concludes that biological intelligence can be considered
as a dimension of information-processing capacity, or behavioural capacity,
that differences among species should correspond to differences in that
overall capacity, and that capacity should be related to brain size
(Jerison, 1978). However, Jerison's view seems to be really a theoretical
guide to the relative intelligence of various animals, and without
confirmation from neuro-anatomical and behavioural studies does not tell
us much about the whale. We prefer the more direct approach of Dr Pilleri
and Dr Morgane, who in concluding that the cetacean brain is 'highly
evolved' and functions on a higher order like that of man, say that that
function must be related to what they term 'intelligent action' in the
context of the whales special aquatic world (Pilleri and Morgane, 1978).

Evolution of the brain

The most ancient known fossil whales, from the Eocene period (about 50
million years ago) were already completely aquatic but had rather small
and primitive brains. They may have evolved from mammalian carnivores which
gradually became adapted to life in the seas. Fossil records of the
earliest whale forms (the Archaeoceti) are fragmentary and limited.
Zeuglodon, a well-known example of a fossilised early whale, had large
eyes, many long pointed teeth, and nostrils at the end of its snout.
Gradually the skull shape of early whales altered, resulting in a
telescoping of the skull and a movement of the nostrils to above the eyes.
In the upper Oligocene period, about 30 million years ago, there first
appeared dolphin-like animals similar to today's river dolphins.
Delphinidae, the modern dolphins, began to appear in the early Miocene
period, about 25 million years ago, and became common in the seas within
the next 10 million years. Sperm whales appeared about the same time.
Fossilised records of baleen whales (Mysticeti) are first found in the
Oligocene period, and appear to have evolved directly from the Archaeoceti,
since skull telescoping occurs in a different way in baleen whales than
in toothed whales (Bunnell, 1974).

The following general statement provides a background to our present topic (Jacobs, Appendix 16):

In all animals with backbones, the brain is the chief organ of the central nervous system (Brain + Spinal Cord). The pre-eminent position of the brain has apparently resulted from processes that have acted on the nervous system of animals throughout evolution. With the appearance and ascendance of mammals in the animal kingdom, the brain has become progressively larger. This has been accomplished primarily by the addition of more neural tissue to the cerebral hemispheres (neocortex) and cerebellum (neocerebellum). It is the great surface mass of neocortex in the human brain, as compared to other land mammals, that is generally accepted as the basis for man's ability to ideate, symbolise, create language and develop implements to manipulate his environment.

The whale is endowed with a neocortex that is even more abundant than that of man and for that matter any other mammal. Despite this common neural basis for intelligence, the extreme dissimilarities of environment and the sound communication barrier have, thus far, stood in the way of general acceptance of the view that whale and human brains share the same functional capabilities.

The typical cetacean brain is globular and is both higher and wider than it is long; in these features it contrasts rather sharply with the brains of most terrestrial mammals. It has a peculiarly compressed appearance and is foreshortened, features which have apparently developed in conjunction with an enlargement of the head region, during prolonged isolation in the sea. The alterations in the area of the head contributed to the gradual movement of the external nasal openings to the forehead region, where they became the whale's blowhole and also, by removing the bony orifices through which the olfactory nerves pass, resulted in the partial, or total (for toothed whales) loss of the sense of smell (Jacobs, 1974).

These developments in the head region and its large size have not only been responsible for the brain's peculiar shape but also are paralleled by a massive enlargement of the brain. In part this appears to result from greatly increased sensory inputs through the cranial nerves, especially the auditory, facial and tactile nerves. Indeed because of the nature of the environment the head region of whales seems to be more important in receiving these influences from the world around than is the case in land animals. The bulk of the increased brain size, however, is the result of an expansion of the two cerebral hemispheres, most of which are covered by neocortex. This is what sets mammals apart from other animals behaviourally and enables them to respond with more adaptability and flexibility to new situations (Jacobs, 1974 and Appendix 16).

One main approach to the study of whale intelligence is to be found in comparative studies of the cetacean brain and the brains of primates, including man. It is generally accepted in anatomical studies that structure cannot be divorced from function, though most of the rules of translation between them are not known. Nevertheless there is every reason to believe that differences in structure mean differences in function. So, to relate and evaluate the status of a brain, scientists use the comparative method of analysis of brain structure in relation to behaviour across many lines of animals at all stages of their evolutionary history. Simpler brains are often related to simpler forms of behaviour and limits in response and activity are assessed in terms of the amount and quality of their parts (Morgane, 1978(a)). As it will be seen that we are unable to come to any definite opinion upon the conclusions reached by this method, we propose to discuss the issues only in a most general way.

First let us turn to the human brain. Each hemisphere may be subdivided into three concentric lobar formations, arching from front to back, that reflect different stages of brain evolution. The innermost and most ancient lobe, lying closest to the midline, is very poorly developed and has been termed the rhinic lobe (from the Greek rhinos, the nose) because it was once thought to be that part of the brain having to do with smell. The next lobe, modest in extent, has been called the limbic lobe (from the Latin limbus, border) because it is disposed along the border of the hemisphere. The third and by far the largest of the lobar formations lies superimposed on the limbic lobe and hence has been termed the supralimbic lobe. This represents the most recently acquired part of the hemisphere and makes up the bulk of the middle portion and those portions towards the back. Further, each hemisphere has a cellular covering or cortex and well over 90 per cent of this surface lies over the newer portions and hence is referred to as neocortex (Jacobs, 1974).

The neocortex or 'new brain' in its appearance may not seem so grand to the layman. Morgane has described the neocortex of the dolphin as 'a wrinkled and convoluted gray layer about one-eighth of an inch thick that covers the surface of the cerebral hemispheres. It is made of nerve cells and innumerable fine, branched and intertwined nerve fibres. It is a place where various kinds of nerve impulses from all over the body are brought together and recombined into outgoing nerve currents' (Morgane, 1974). The processing of the sensory information received in the brain, the direction of motor activity of the body, the control of muscular movement, and indeed all the functions of the brain including the higher ones depend on the organisation of the nerve cells, and the transmission of bioelectrical impulses through their pathways. So the branching and linking of the neurons and the bioelectrical circuits also determine the brain's quality (Jacobs, Appendix 16).

A major similarity between the cetacean brain and the human brain, as we have noted, is the very great expanse of the neocortex or 'grey matter'. Observations upon the dolphin and other species suggest that this is a general feature of the cetacean brain (Jacobs, 1972). As Morgane says, '...neocortex is the single greatest factor in brains that determines their rank as "highly evolved" and thus brains with high quality neocortex are capable of organising outputs that we broadly term "intelligent" behaviour. In this regard only the brains of whales and men have the amount and quality of neocortex making both appear at the pinnacle of the animal kingdom in terms of our modern views of criteria of brain status' (Morgane, 1978(b)).

The expansion of the neocortex in the primates, including man, has been due to the elaboration of existing cortical areas in lower mammalian forms, and also the development of new ones with the high order functions already noted. For example, the increase in the sensory (vision, hearing, taste and touch) areas provides for increased discrimination and richer perception. The new areas have been developed between the sensory areas that exist in lower forms. These new areas are most developed in man, and provide for the closer correlation and more complete integration of all types of bioelectrical impulses. To distinguish them from the correlation areas which are specifically related to the different sensory reception areas, these new higher general correlation areas are known as association areas (Morgane, 1978(a)).

This brings us to another major similarity between man and whales. The massive associational areas of whale brains are one of their most striking features. Elaboration of these association areas increases the number of choices an animal can make and this is related to what is called 'intelligence'. In the whale brain not only does this elaboration occur, but it is accompanied by a characteristic found also in man and other primates: the visual, auditory and sensory-motor regions occupy a relatively smaller proportion of the brain in the whale than in lower animals, where they take up most of the cortex (Morgane, 1978(b)). In the cetacean brain the association areas are found in a well-developed additional lobe between the limbic and supralimbic lobes, which has been designated the paralimbic lobe. Although the functional significance of this feature (which is not found in the human brain) is not fully known, it appears to be a continuum of all the specific sensory and motor areas which, in man, are distributed throughout the supralimbic lobe (Morgane, 1978(b); Jacobs, 1974).

Turning briefly to the internal organisation of the neocortex, we found that the first feature is in fact also related to its size. This is the degree of folding of the cortical surface, or 'fissurisation', which is luxuriant in cetacean brains, and adds to the great surface area.

Next, in the human brain the neocortical formations are layered (laminated) structures consisting of six layers of cells separated, in various degrees, by fibres, and in higher mammals they are differentiated into many sub-areas of neocortex. The more areal differentiation, theoretically, the more types of function are represented in the cortex. The 52 different patterns of cellular organisations which are referred to as Brodman's cortical areas, and many more, have been distinguished in man's brain by tissue analysis. Both these features of lamination and the degree of areal differentiation have been accepted as criteria for higher development of the brain (Morgane, 1978(b)).

When we turn to the cetacean brain - and most studies have been conducted on the bottlenose dolphin - we find that the neocortex is differentiated into regions or areas in a basically similar way to the neocortex of other mammals, including man, although the boundaries are much less well-defined than in land-living animals. Using modern staining techniques Morgane has so far clearly identified some 30 to 40 types of neocortical area, although no 'brain map' has yet been completed (Morgane, 1978(b)). Further the fundamental plan of the six-layered cortex is the same as in all mammals, but in their individual make-up the layers show special characteristics not found in land-living mammals (Morgane and Jacobs, 1972). There are other differences, including the paralimbic lobe, and also in the complexity of the cellular branching and interlocking (Morgane and Jacobs, 1972). However the similarities may be, if anything, more important than the differences. It is a striking feature that no basically new structures are found in man's brain that are not found in the higher primates, and that similar structures are found in the whale with its separate evolution.

Morgane and Jacobs (1972) point out that 'Obviously, functional interpretations are highly speculative pending physiological and behavioural studies and these latter are extremely difficult to carry out in the cetacea, especially the larger ones'. In fact with the exception of a motor 'mapping' study by Lende and Akdikmen (1968) there has been almost a total absence of these studies in the classification of structures in the cetacean brain. This position has not changed (Inquiry Hearings, pp.708, 709, 714). But the Inquiry appreciates that if scientists have to await the final confirmation which can be provided only by these tests, any conclusion is a long way off.

In summing up their position Morgane and Jacobs took the view, which Morgane reiterated in 1974, that the quantity and quality of grey matter in the brain is to be taken as a rough index of the relative efficacy of the brain in controlling behaviour. They concluded that there was a real question about whether the quality of the cetacean cortex was comparable to that of the higher primates. In commenting on this passage, Dr Jacobs said they were being very general, 'admitting the possibility that since there is very little information available about the cetacean brain, that all of the facts are not in...(It was not) meant to imply that this animal does not have intelligence based on the neocortex' (Inquiry Hearings, p.902). This is the view expressed in his summary that 'although there is a difference in the distribution and pattern of the organisation of neocortex in the whale, the massive volume of it is identical or similar to that of man, and since its cellular organisation is so similar I have no way of saying that functionally it should be any different. The impression I get is that function potential is just as great in the whale as in man' (Inquiry Hearings, p.704). Although most of the work has been done on the brain of the bottlenose dolphin, the basic similarity so far found in the brains of other cetacea has been taken to be sufficient for this view to be applicable generally to cetacea - particularly to the toothed group of whales (Jacobs, Inquiry Hearings p.725) - although significant variations between the species are likely to occur, as in the case of the higher and lower primates.

Professor Harrison, Professor of Anatomy at Cambridge University in the UK, who is a prominent authority on marine mammals and editor of Functional Anatomy of Marine Mammals in which the joint paper by Morgane and Jacobs is included, takes a much more conservative view. Harrison's impression as an anatomist is that the brain structure is typically mammalian, with nothing especially unique, but reflecting development of some parts which are not correspondingly developed in the human brain. Generally there is very little known about differentiation and function of cetacean brains, he considers, especially in the case of large whales, which pose extreme practical problems for experimentation. Harrison does not believe that cetacea are intelligent, and while 'intelligence' in relation to cetacea is a vague concept, some form of intelligence would be suspected, he considers, if the cellular density in the cortex and degree of connectivity were similar to man's and supported by certain behavioural patterns (Harrison, 1978).

The Inquiry has considered the views on cetacean brains expressed by these scientists on the basis of comparative studies. No information is available as to any other studies. The field is a specialised one, and we would hesitate to draw final conclusions without a sound basis of scientific proof. So we are unable to go all the way and find that the whale is an animal with intelligence similar to that of man or of a high intelligence potential. However, because of the striking features which do exist in the cetacean brain, we consider that the issue should be left open and that there is a real possibility that this potential does exist and, in the words used by Morgane in 1974, that 'we are dealing with special creatures with remarkably developed brains'.

Brain size as an indicator of intelligence

This brings us to Jerison's paper, where the emphasis is on the great size of the cetacean brain. The usual criticism of gross brain size as an indicator of intelligence, says Jerison, takes the form of statements such as 'its not the size of the brain that counts, rather it's the complexity of its convolutions'. More sophisticated criticisms might claim 'brain size is trivial. The important thing is the pattern and complexity of interconnections of nerve cells'. Jerison considers these criticisms invalid. He says we actually know more about brain size that is useful for comparisons among different species of mammals than we do about any of the microscopic measures (Jerison, 1978).

Jerison points out that all mammal brains can be thought of as bags filled with neurons, glial or supporting cells and all the other nervous tissue components. As he says, a bigger bag holds more cells, and the number of neurons in a mammalian brain can be estimated from brain size. As a second approximation one must take into account the organisation of nervous systems into networks of neurons that are in contact with one another. The amount of communication in all mammalian brains is enormous. There are about 10,000 million neurons in the human cerebral cortex and 30,000 million in some whales.

The weights of the brains of various cetacea and man have often been compared. The following table, based on data by Morgane and Jacobs (1972), Bunnell (1978), Pilleri (1975) and Lilly (pers. comm., 1978) gives an indication of the range of sizes. The massive size of the sperm whale's brain is particularly noteworthy. The brain of the di[...] tiny compared with that of the whale.

Ganges River dolphin	250 gms
Harbour porpoise	400 gms
Human	950-1700 gms
Bottlenose dolphin	1600-1900 gms
Pilot whale	2500-3000 gms
Beluga	2500-3000 gms
Fin whale	5000 gms (approx.)
Blue whale	6000 gms (approx.)
Killer whale	6000 gms (approx.)
Humpback whale	6000-7000 gms
Sperm whale	8000-9200 gms

Jacobs (1972 and 1978) states that the smallest brain weight for a human brain believed to be compatible with normal 'intelligence', and considered to be critical for supporting a language function is between 900 and 950 gms.

The central point in Jerison's analysis is that the brain is an organ of the body, and animals with bigger bodies should be expected to have larger brains - just as their other organs such as the liver, heart, lungs and skin are bigger - because there is more body for the brain to control. The next step is to divide the size of the brain into two factors, one related to body size, and the other a 'residual' factor of brain size related to an information-processing capacity beyond the minimum required to control the body. It is this second factor that Jerison labels as an 'encephalisation' factor, so that a higher factor corresponds to brains of a higher functional or, let us say, intellectual capacity. Jerison concedes that the analysis is not conventional.

Jerison, building on data of gross brain and body weights, shows that the present cetacean grade of encephalisation overlaps those of the most highly encephalised primates. Human beings are shown of course at the top, with various grades of encephalisation in the various primates. Jerison shows that various grades exist also in living cetacea with the bottlenose dolphin at the top.

eds to deal with the interpretation of encephalisation.
sual auditory and vocal adaptations in the whale
d as a possible explanation for the expansion of
easoning, he says, runs into problems because
daptations in bats which also have echo-location
ained mammals. Jerison comments that the increased
ing machinery in a whale's brain compared to a bat's is
that it is simply unrealistic to assume that an unusual capacity
for echo-location and ranging is enough of a drain on a nervous system to
require the enormous extra amounts of tissue. He says it makes much better
sense to assume that cetacea take the unusual information from sound
ranging and echo-location and do other unusual things with it, about which
we know little or nothing (Jerison, 1978).

Jerison says it is reasonable to assume that human intelligence has
counterparts in other species. So he concludes that if intelligence is
defined as encephalisation then human beings are to be considered as
part of a set that also includes some cetacean species, particularly the
bottlenose dolphin and the killer whale, and to an unknown extent (since
the concept of encephalisation runs into difficulties with the truly giant
species such as the sperm whale), the very large whales.

There is of course no scientific proof for Jerison's view. Indeed
there is no general acceptance of the significance of the brain weight or
brain to body ratios with respect to the higher development of any given
brain (Morgane and Jacobs, 1972; Morgane, 1978(a)). Jacobs also points
out that much depends on the selection of the species. It would be
necessary to exclude brains of primitive or lower mammals where the weight
is not so much related to the neocortex (Inquiry Hearings, p.758).
However while on the present state of knowledge the Inquiry cannot draw
any conclusion about intelligence from Jerison's paper, perhaps the
approach does provide some support for the views of the anatomists, for
as Jerison says, where there is information about gross brain size and
microscopic measures we find remarkably orderly relations between the two.

Communications systems, sonar and echo-location

This brings us to a feature which also has bearing upon the whale's intelligence.

Aristotle was aware that dolphins could produce sounds at the surface of the water. But a belief that the large whales were silent persisted until the end of the eighteenth century when the screams of wounded whales were reported (Slijper, 1962). Now, of course, the songs of the humpback whale are familiar to many from the disc recordings by Dr Roger Payne.

Sound is the primary sense for cetacea. Vision is often difficult in the dark and murky seas and impossible at great depths. There are many species which navigate and hunt at night or below the zone of illuminated water. Sound travels extremely well in water and can be used at any time of the day or night and at all depths. There is little doubt that whales, as herd animals, communicate and locate with sound and can keep in contact by sound quite easily. Cetacea have evolved at least two methods of producing sounds, pure tone or modulated whistles being used predominantly for social communication, and click sequences for navigation and for the location of submerged objects by interpreting echoes. This second type of voice is often referred to as 'sonar'. The two voices overlap in both quality and function (Warshall, 1974). (Echo-location is not, of course, unique to the whale - it is found in bats and certain seals.)

Although whales have no vocal chords they have a well-developed system of sound generation through special muscle arrangements and air cavities within their bodies which are used to produce sounds (Pilleri, 1971). The movement of trapped air between air sacs is responsible for a wide variety of sounds emitted by cetacea (Jacobs, Appendix 16).

Dr Lilly has studied dolphins for many years and has concentrated to a large degree on acoustics, echo-location and sound production. We can make only a most general reference to his findings. Dolphins, says Lilly, live under water and are not able to make use of the atmosphere in emitting sounds. They have three sonic or ultrasonic emitters, two of which are in the nasal region on each side of and just below the blow hole, and project sounds in all directions around the dolphin's body. The third is in the larynx and enables the dolphin to emit pulses of sound, focussed in a narrow beam, of a higher frequency than the sounds produced by other means. The rate of these pulses can be controlled from one per minute up to one thousand per second. This is the output side of the dolphin's so-called sonar, sonic navigation and ranging system. The pulses go out through the water, are reflected by objects, and come back to the ears of the dolphin under water. While the dolphin is using this system it moves its head horizontally to scan any object with the tight beam of pulses. Returning pulses are received by the dolphin's ears, and then interpreted by the brain. The dolphin has been shown to discriminate objects hidden from its eyes, with exquisite fineness, by the use of this system (Lilly, 1978).

There is no doubt that the sense of hearing is very acute in toothed whales, and that these animals, including the sperm whale, are capable of detecting and analysing minute amounts of sound (Gaskin, 1976). The sounds used by the sperm whale are predominantly low pitched for long range deep water search. Watkins (1977), who conducted experiments at sea on sperm whale sounds, was convinced that in their underwater movements the whales reacted to each other, both acoustically and physically, and it may be that all sperm whale sounds have a communicative function, such as maintaining contact with others in a widespread herd. Watkins considers that, while at present too little is known for certainty, it is obvious that sound plays an important part in whale behaviour.

Researchers, in particular Lilly and Payne, have concentrated special attention on the communication abilities of cetacea. Lilly (1976) says: 'It is highly probable that their minds operate with acoustic analogs, even as our language is primarily based on visual analogs...They can see emotional states in one another through stomach and lung movements. They become acquainted with, and subsequently recognise, one another and humans in the water with them, through the transmission of sonic pictures'. It should be noted that there seems to be only behavioural evidence for this degree of sophistication in the use of sonar (Inquiry Hearings, Greenwood, p.887). In Lilly's experience (Lilly, 1976) dolphins are sufficiently interested in communicating with man for them to make surprising efforts to 'reprogram' - to speak in air and mimic human speech - and solve communication tasks imposed by man.

Lilly (1977) says that the fundamental requirement for testing cetacean intelligence is to establish a means of direct communication with dolphin or whale. Scientists are trying to communicate with other animal species and the work of Premack, Gardener and Rumbaugh, for example, on communication with various primates by the use of symbolic devices has revealed many new areas of their behaviour and reputed intelligence. At Lilly's Human/Dolphin Foundation in Malibu, California, a program is being developed using high speed computers to establish the physical sonic doorway required in communication between man and cetacea to allow for differences in the frequencies available for sound and hearing.

Lilly is aware that many regard certain of his high expectations of dolphins as science fiction. He is satisfied, in scientific and humanistic terms, of the probability that cetacea are sentient, intelligent creatures, and speculates that they have mental activities based on memories which may be greater than man's. He accepts the neurological evidence and relies also on the behaviour of cetacea, including evidence of an ethic among the dolphins which assigns to man a 'very special station', and also their group solidarity. He instances the group care of the sick or injured dolphin, but points out that if it 'interferes with group survival, the individual voluntarily stops breathing and thus commits suicide'. Finally, he points to the sonic/ultrasonic communication system which is more complex, faster and ten times the frequency of man's (Lilly, 1978).

We make one comment upon Lilly's work. We feel that the fact that only in recent years has the possibility been established of communicating with apes using logical symbols or signs from AMESLAN (American Sign Language for the Deaf) in place of words (Jerison, 1978), should lead us at least to withhold judgment upon Lilly's attempts to establish communication with cetacea.

Dr Roger Payne has been involved for over twenty years in research on cetacean acoustics and behaviour. He says that whales make various sounds, which possibly function as echo-location sounds, social sounds or songs (Payne, 1978). Payne's experience has been with humpback and right whales. Humpbacks in particular have well defined 'songs' lasting from about six to thirty minutes and then repeated for many hours. Payne's research currently indicates that only adult males sing. He says that all individuals in an area sing the same song but individual whales can be identified in the same way that human voices can be distinguished. Different whale songs occur in different areas and apparently the song changes during a year. A series of Hawaiian observations indicates that whales start singing each year with the song that finished the previous year. The songs are gradually modified during the year and are produced during the breeding migration and in the breeding area. Whales do not sing in the feeding area; vocalisations made there are 'social' and similar around the world. Changes to whale songs are not random, Payne says, but have evolved according to a set of 'laws' similar to the rules of musical compositions. Payne's observations show similar results in Hawaii and Bermuda. The laws may be inherited culturally or genetically. It is most unlikely that mixing would occur as the whale stocks are isolated. Payne says that the whales' capacity to remember a song during the feeding season must require a substantial 'storing' ability in the brain.

Sounds produced by sperm whales seem to Payne to be echo-locating and social sounds, but not songs. Direct extrapolation of observations upon humpback whales to sperm whales is not appropriate, according to Payne, and only general parallels can be drawn.

In Payne's opinion a display as elaborate as these songs is unlikely to be trivial, and whales' large brains may in part be related to the need for substantial storing of information. He feels that the relevance of the size and structure of the sperm whale's brain is still unclear and that comparison of its association areas with those of other animals is a matter for caution as whales have different association requirements. On the whole he says it is not possible as yet to compare brains very effectively.

Although the meaning of the sounds made by the various species of whales is still largely a matter of speculation, the sophistication of cetacean use of sound is generally considered to indicate at least a brain with a highly developed ability to store and process information.

Behaviour and intelligence

Whales are known to exhibit varying degrees of social organisation within their populations (see Chapter 1) and this may be evidence of a sophisticated level of behaviour and biological intelligence. This view is strengthened by the consideration that juvenile whales spend long periods with their parents and other adults, and, in particular with the sperm whale, have ample time to learn from their seniors and to be 'taught' certain basic and vital behaviour. Professor Singer, in his submission to the Inquiry, points out that the whale is an 'intelligent, social species where emotional links between different members of the group, and the capacity to enjoy life, are only too evident' (Singer, 1978).

The significance of an advanced social structure to the level of 'intelligence' of the particular animal is not fully understood, and it is well known that other animals, ranging from bees to wolves and gorillas, have a developed social structure. Nevertheless, it is apparent that whales have developed this system over millions of years, and that in some ways at least it may mirror several aspects of a social community found in man.

The 'epimeletic', or 'care-giving' behaviour of cetacea is sometimes instanced as an illustration of an advanced social system. Scientific awareness of epimeletic behaviour dates as far back as Aristotle, who remarked of the dolphin that the creature is remarkable for the strength of its parental affection. He also related a tale of the wounding and capture of a dolphin, when the remainder of the school came into and stayed in the harbour until the fishermen let their captive go free, whereupon the school went away.

Caldwell and Caldwell (1966) summarised the evidence of epimeletic behaviour for cetacean species. Several species are known to stand by when a young of their species is injured, while others also stand by injured or distressed adults. Certain species may support injured animals at the surface. In some of the species that stand by, including the sperm whale, other responses that apparently result from excitement have been reported. Circling rapidly around the distressed individual is common. Sometimes those standing by have broken harpoon lines, pushed the injured animal away from capturing boats, or inflicted damage on boats. On occasion, they may position themselves between the injured animal and the attacker.

Caldwell and Caldwell point out that sexual differences may exist in the behavioural responses. They say female sperm whales tend to remain with distressed females or young, but usually leave males in distress. Male sperm whales rarely, if ever, remain with distressed females. In contrast, certain female dolphins support injured males.

Caldwell and Caldwell comment that in assessing these observations it should be remembered that certain cetacea have been widely hunted while others have not been hunted at all. Consequently it cannot be assumed that a behaviour pattern does not exist in a species merely because it has not been reported.

Looking at cetacean behaviour more generally, anecdotal evidence of sophisticated behaviour by certain cetacean species is widespread in the literature, with such famous novels as Moby Dick introducing the reader to the many different behaviour patterns of the sperm whale. Other early accounts refer to certain species of dolphins and porpoises helping local communities in some areas of the world to catch fish (see ACMRR/MM/SC/5 Add. 1,1976).

One example of this is provided by the Australian Aborigines, in particular the tribes on Mornington Island, who are believed to go through special religious performances calling the dolphins in from the sea to help the tribe in their fishing efforts. This is described by Project Jonah:

> Jackson Jacobs, a tribal leader from Mornington Island, who is of the dolphin totem, recounted in a taped interview how he used to watch his father and grandfather fish with the aid of the dolphins. He sang a song that his father used to sing, which included a

high-pitched quivering whistle, and he also mentioned how his father would clap in a certain way under the water and slap his spear on the water. One or two dolphins would then appear and drive the fish in towards the beach where his father could catch them. When sufficient had been obtained, the best fish would be selected out and thrown back to the dolphins as a token of thanks (Project Jonah, 1978, p.3B/36).
In this and other accounts one motivation of the dolphins appears to be the satisfaction of their own needs.

There are many more recent examples of interesting and novel behaviour in cetacean species, usually dolphins and porpoises, but also including pilot whales and killer whales. Project Jonah has mentioned behaviour which seems indicative of a high level of cerebral involvement including the ability to play and enjoy games not obviously related to hunting skills (instances are given both in aquariums and, on the part of the great whales, in the open sea), the ability to initiate ideas (such as new tricks), expressions of grief, joy and sense of humour, the use of sexuality in a social role, not just for reproduction, and co-operation with human beings (Project Jonah, 1978, pp.3B/125-47). The fact which appears from one of Lilly's experiments, that it took only a short time to persuade a dolphin to wear suction cups over its eyes in order to test its echo-location system, or the fact that a dolphin would bother to dive 1000 metres in the ocean to press a lever to take a photograph of itself, suggests a brain functioning at a sophisticated level (Project Jonah, 1978, p.3B/36). Some instances have been cited where it can be suggested that cetacea are capable of abstract thought or foreseeing the consequences of actions, or the communication of these aspects, which have been suggested as tests of behaviour at a level similar to man's. However, the strict application of these tests has also been questioned, and it has been pointed out that much human communication is not utilitarian in the sense that it is necessary or adaptive (Project Jonah, 1978, p.3B/68).

Mr M. Greenwood presented evidence to the Inquiry on his research while he was a scientific adviser working for the United States Government on marine mammal weapons systems. Mr Greenwood's research on cetacea was conducted using Atlantic bottlenose dolphins, pilot whales and killer whales.

The United States Navy is attempting to determine what characteristics these marine mammals exhibit in their natural environment that might have broader application and use. Research has included deep diving studies to learn more about the physiology of marine mammals and the possible application of those findings to navy personnel and others who have to venture into the oceans. Other areas of study being undertaken by the United States Navy include the movement of marine mammals through water, their speed and endurance and their sonar systems. Benefits from these studies have been applied in submarine and sonar design. According to United States government information, this research over the years has developed new knowledge and techniques for preserving marine mammals. Many of the specific ideas on the care and handling of marine mammals developed in implementing the Marine Mammal Protection Act of 1972 is based on this research by the Navy.

Official United States information shows that trained pilot whales, killer whales, and porpoises, together with California sea lions, are able to assist in finding and retrieving from the deep sea, test weapons containing acoustic beacons (and indeed on Greenwood's evidence (Inquiry Hearings, pp.863-4) objects not fitted with such devices). Dolphins have also been used to help man underwater as in the United States Navy's Sealab Project of several years ago. Dolphins were trained to carry tools to divers and to help find lost divers.

In his evidence Mr Greenwood mentioned the killer whale and its placid and gentle behaviour towards man, with none of the attacks or aggressiveness which it certainly shows to other species. (Its name is derived from its practice of preying on other whale species.) Mr Greenwood commented that there is a discipline and control displayed in the killer whale that is extraordinary. He said that under conditions of extreme provocation the killer whale is able to control its own emotion and does not revert to a sort of reflex behaviour or attack, as would many other animals, including man.

It is important to note that the experimental work has been done in the open sea where obviously the whales have the option of ignoring the researchers and swimming away. In fact the animals stay with the research workers' boats for reasons which would appear to be related only to their fascination in the task in hand. Some patterns of behaviour recounted by Greenwood are remarkable, and would seem to justify his claim that the cetacean brain operates in a quite extraordinary way, and that the animal is capable of patterns not seen in other animals. However he says that it probably has a different type of intelligence from man's, though not necessarily inferior (Inquiry Hearings, pp.876-7, 890).

Karen Pryor has concentrated on observing and training small cetacea in captivity and has described many examples of novel behaviour and advanced learning in captive specimens (Pryor, 1969 and 1973). She has placed the smaller cetacea into three definable groups, correlating with feeding habits (Pryor, 1978). The first group consists of the beluga and pilot whales and the Amazon dolphin which feed predominantly on squid and organisms that live on the ocean bed, and take their food where they find it without having to chase it. These species demonstrate little or no aerial display, minimal interest in objects or playing with objects and low levels of aggression.

A second group includes dolphins which feed on fast-swimming prey that must be chased. These coursing predators demonstrate high aerial activity and some forms of play, but are afraid of objects, have no skills in manipulating them, and are very group-dependent.

The third group consists of the false killer whale, the killer whale and the rough-toothed dolphin. They were found to be very good at manipulating objects and able to tolerate solitude well. They were also very playful with objects, but very aggressive to each other, to people and to some objects (for example when in captivity they often smashed apparatus). This group feeds on large prey which must, at least on occasion, be fought and subdued. Both species of bottlenose dolphin were anomalous in that they seemed to fall between the second and third group.

Pryor considers that intelligence is certainly not lacking in any of these species, with certain species demonstrating varying degrees of rigidity and plasticity in their behaviour (e.g. studying latch gates, opening gates, problem solving).

Pryor endeavours to place whales in similar categories to the smaller cetacea which have been studied more thoroughly. Whales which feed on food that is easily found, and which travel slowly and are not predatory (for example gray whales), probably resemble the first group behaviourally (and in 'intelligence' therefore), and would be poor problem-solvers, showing little aggression and remaining fairly rigid in their behaviour, not adapting well to change.

Whales which feed on large prey which must be hunted and subdued should resemble the killer whale group, showing active aggression to organisms and objects (for example by attacking the whalers' boats) and should exhibit novel behaviour and the capacity for problem-solving. So far as the sperm whale is concerned the historical evidence of early whale-chasers seems to fit in with this suggestion.

Pryor suggests that there should be a body of whales in between which are more streamlined than gray whales, and feed on faster-moving prey, resembling in temperament and levels of behavioural plasticity the middle group of dolphins. Perhaps, as Pryor suggests, the blue whale fits in here.

Pryor concludes that all the best behavioural work is so new and so sparse that there is very little conclusive evidence on which to base scientific opinion. Barring neurological speculations she sees no evidence to suggest that any of the whales are 'smarter' than chimpanzees, people, or the 'smart' small cetacea.

In summary, Pryor says that few whales are, or need to be, as opportunistic as the bottlenose dolphins or as sharp at problem-solving as the killer whale. Therefore, whales are by and large 'no geniuses' (Pryor, 1978).

One argument commonly put forward to show that whales are not intelligent is that they continue to come in large groups to the whaling grounds, such as those off Albany. It may be of course that they are driven there for food. A contrary view on their intelligence has been expressed by one of the Albany catcher skippers: 'Oh, they're intelligent alright. They know when a ship's after them. They dart this way and that and then submerge. One old bull we timed went down and stopped on the bottom for three hours' (which is very much longer than whales usually spend submerged) (Hutton, 1978).

Some scientists also remain sceptical about the intelligence of whales for other reasons. Harrison (1978) comments that if cetacean behaviour exhibited features such as an ability to count, an ability to respond to a series of commands before a sequence of tasks, or an ability to contrive escape from holding facilities then one might suspect that cetacea were 'intelligent'. He considers that most behaviour of **cetacea** in captivity reflects natural behaviour, for example, jumping, fetching and diving, and is a response to food rewards. Harrison believes that the cetacean brain is not comparable to man's and that while, for example, dolphins may be trained to fetch identified objects or to respond to their individual names, in these respects they are no more capable than a well-trained sheep dog. He says that attempts to train dolphins for activities such as pipeline monitoring or for military purposes have not been encouraging. But the evidence mentioned above from United States Government studies is very much to the contrary.

Jerison (1978) refers to the subject of formal behavioural tests of 'intelligence' with cetacea. He asserts that these behavioral tests are in their infancy and involve almost simple-minded applications of a few tests that have been used in this way in primate species. Jerison says it is not surprising that the dolphin did well in such tests, but it is also not surprising that the dolphin did not surprise its tester by its ability. We have not discovered how to communicate with dolphins and whales. Perhaps, says Jerison (and Lilly) we will, since it is only recently that man has started to learn to communicate effectively with the great apes. Jerison also says that in his view plasticity of behaviour or ability to learn are weak measures of an animal's 'intelligence', and that all animals are able to learn as much as they need.

As is evident from the various views presented on the whole issue of whale intelligence and behaviour, it is difficult to make precise and factual statements. As Dr David Caldwell (1978) from the University of Florida states:

I am sorry to say that we can really say very little about whale intelligence...We have essentially no experience with the large whales, but people tell us that they _feel_ that they may be a bit more sluggish (except for the sperm whale) than the dolphins. No real basis for this, however, and it just may be that people feel that way because the large whales are generally slower in their actions and less 'playful'.

As far as we know, there just aren't any good experiments that deal with the matter of cetacean intelligence. It would be a hard thing to test in comparison to other mammals. And in fact, I think a lot of scientists aren't really sure what 'intelligence' is in humans. It probably really doesn't matter when one considers that the real problem for an animal is how to survive in its particular environment and not how 'intelligent' it is compared to some other kind of animal that may be doing perfectly well under the circumstances of its own environment.

Conclusions

We consider that there is much force in the summary statements made by Dr Myron Jacobs (Appendix 16):

> Three facts are of cardinal importance when comparing the functional capacity of the whale brain to that of man. First this magnificent line of mammals has been evolving for a considerably longer period of time than man. Second, adaptations and specialisations occurring during prolonged isolation in the aquatic environment have resulted in a body form that is entirely different from man's, one that is reflected in adjustments in the organisation of the central nervous system. Third, despite the prolonged separation from terrestrial lines of mammalian evolution, whales have brains that are quite comparable in size and complexity to that of man, a state not achieved by any other aquatic mammal or, in terms of size, by any primate. In view of these facts, whales as a group should be thought of as having reached a level of morphological and functional brain development in the aquatic environment comparable to man's in his environment.

However, as we understand Jacobs' conclusion, we feel that some qualification must be made on the positive assertion of the whale's functional brain development. We have already indicated that on the neuro-anatomical evidence the Inquiry is unable to make the assumption of a potential for high intelligence in the whale. But we are persuaded by the evidence submitted to us that the issue remains open and there is a real possibility that such a potential exists, and that accordingly, allowance for it should be made in man's attitudes to whales.

Certain whale species, particularly some dolphin species and the killer whale give evidence of advanced behavioural activities. It is from these behavioural studies that scientists have endeavoured to draw parallels for other whale species. Granted that many assumptions have had to be made, nevertheless it is not unreasonable to conclude that cetacea give evidence of levels of behaviour that would seem to be associated with a level of brain development and activity of some sophistication.

But the whale remains a mystery, and it seems that no final solution will be found until there is a scientific break-through of the first magnitude, enabling the sounds by which the whale communicates to be understood. This seems a long way off.

10. TECHNIQUES USED TO KILL WHALES

What an outcry there would be if we hunted elephants with explosive harpoons fired from a tank and then played the wounded beasts upon a line.

This view of Professor Sir Alistair Hardy, zoologist to the <u>Discovery</u> whaling expedition to the Antarctic in the earlier part of this century, was cited by Project Jonah (1978, p.3D/22). A similar view, using a bullock for comparison, was attributed to the late Ludwig Glauert, distinguished biologist and museum director, by Mr Vincent Serventy, President of the Wild Life Preservation Society of Australia. Serventy, who has himself been out twice on whaling ships, adds: 'If the whale had an audible voice and more Australians saw the brutality of its slaughter, whaling would be stopped' (Wild Life Preservation Society of Australia, 1978, p.5).

The view that the methods presently used to take whales are inhumane has been supported in many submissions to the Inquiry, including those by the International Society for the Protection of Animals and several Australian branches of the Royal Society for the Prevention of Cruelty to Animals. Not all these submissions argued for the absolute protection of whales; for example, the Royal Society for the Prevention of Cruelty to to Animals, Western Australia said on page 1 of its submission 'whaling should be discontinued unless a more humane method of slaughter be introduced'.

The claim that whaling is cruel thus has widespread and authoritative support. It requires due consideration.

History of past techniques

Techniques used to hunt whales have a long history, with methods varying according to the locality and cultures of the hunters involved and the species of whales captured.[1] At least three different methods were developed for killing the large whales; trapping with nets in embayments and killing the trapped whale with poison arrows, hurling hand-held poison-tipped harpoons into surfacing whales on the open sea, and open-boat hunting with hand harpoons and attached floats, using a lance to kill the whale.

The earliest harpoon used was shaped like an arrowhead. It was called the 'two-flued iron'. This was replaced by a 'one-flued iron' with a single, fixed barb, and later a harpoon with a single moveable toggle point or barb was used. Modern whaling began in the 1870s with the development by Foyn of the cannon-fired harpoon. Although the technology has improved somewhat since then, the method used, as described later in this chapter, has changed little.

Another method which was tried was the use of hydrocyanic (prussic) acid, placed in bottles in slots in the harpoon head. The bottles broke when the harpoon penetrated the whale, releasing the acid, which killed the whale quickly and efficiently. Unfortunately it also caused the death of some whalers who handled poisoned blubber, and this led to the rapid abandonment of this technique. Other lethal or narcotising drugs have also been used. A mixture of strychnine and curare shot from a hand-held harpoon gun was the subject of experiments by the French investigator Thiercelin in 1886, but although promising the technique was not practised extensively.

The use of electrical charges to kill whales has a long history dating back to the 1850s. One type involved a wire conductor in a hemp whale line fastened to a standard hand-held harpoon. The whale was electrocuted by a current which ran from a hand-operated induction machine to the harpoon.

Clarke (1952) gives a comprehensive discussion of whaling by electrocution. Death times and the current varied depending on where the harpoon lodged in the whale, from 70-80 amps in the heart and lungs with

---

[1] This section draws heavily on Mitchell and Stawski (1978).

instantaneous death, to 35-60 amps in the back muscle with death occurring in ten seconds to two minutes. (Other sources record death times of two to five minutes. See for example Best, 1975.) Clarke comments that using the electrical system for killing whales is technologically more efficient in many ways than the explosive grenade system, but that, in view of the dangers of electrocution, its continued use ultimately depends on the attitude of the whale-gunners.

However, the electric harpoon did not live up to its promise and the problems associated with this method were too great for it to be pursued (Ash, 1964). Research by the Japanese whaling industry in the 1950s found that in fact electric harpoons were less than 50 per cent as effective as the explosive harpoon. Another disadvantage was the discoloration of meat resulting from the burning of muscles. So the method was abandoned. Research was continued, however, on the use of electricity to despatch minke whales which were not killed by the first harpoon shot. The method was for the animals to be towed alongside the catcher boat, electric probes implanted and an electric current turned on. This method is currently in use (IWC/SC/29/Rep. 5, 1978).

Present techniques and future possibilities

Open boat, hand-harpooning methods are still practised in three areas: the Azores for sperm whales, Tonga and Bequia in the West Indies for humpbacks. Bowhead whales are killed from open boats by Alaskan Eskimos using bomb lances and shoulder guns designed during the nineteenth century.

All commercial whaling operations use harpoon cannons. A 90 mm cannon firing a 55 kg harpoon with a cast-iron explosive head is generally used for large whales such as the sperm whale, while 75 mm or 50 mm cannons and smaller harpoons are more commonly used for sei and minke whales. A non-explosive head is usually used for minke whales, to conserve more meat for human consumption (IWC/SC/29/Rep. 5, 1978).

Cheynes Beach uses this orthodox method. The procedures followed are described in more detail by the company:

The method used for taking whales is to locate the animals by the use of a spotter aircraft which searches the ocean surface for indications of the animals or alternatively by direct sighting from whale Chasers (being vessels of some 500 tonnes and 155 feet in length with a single steam driven engine of approximately 1850 hp). Once the animal or animals are located the Chasers move up to them and if the animals should submerge a fishing type sonar device is used by the Chaser to keep contact with the animal and also the aircraft flying above can assist in directing the Chasers to the animal.

When the Chaser is at sufficiently close range the gunner must assess the animal to see that it is within the regulations as demanded through the International Whaling Commission and in turn the Australian Government licensing requirements and then if the animal is within those requirements (in the gunners' judgement) the gunners fire a harpoon at the animal.

The harpoon is fired from a 90 mm cannon and the harpoon itself carries a 20 pound grenade which is controlled by a time fuse and explodes in the animal after the harpoon has entered.

In the majority of cases the animal dies instantaneously but if the shot is not completely effective in killing the animal, a further harpoon known as a killer harpoon is fired. The killer harpoon is in all respects similar to the original harpoon or first harpoon used except that it does not have flukes (hinged steel barbs for holding the harpoon in the animal). The killer harpoon has no forerunner (that is a line running from the ship to the harpoon) (Cheynes Beach, 1978, pp.29-30).

Concern about whaling techniques currently in use led to consideration by the International Whaling Commission and individual nations of alternative methods of killing whales with the aim of developing more efficient and humane practices.[1] The most recent review was at the June 1978 meeting when a subcommittee of the Scientific Committee reported on humane killing techniques.

---

(1) See for example the 1977 reports of the Scientific Committee and the Commission (IWC, 1978) and IWC/SC/29/Rep. 5 (1978).

In relation to the use of drugs to kill whales, the subcommittee said that these would 'undoubtedly be effective but the problems of calculating dose rates, delivery methods, drug residues in meat destined for human consumption, and the possible dangers to gunners and crew involved in handling equipment render this method impractical for the present, although if some of these problems can be overcome the technique is promising' (IWC/30/4, Annex J, 1978, p.2).

The subcommittee noted advances in the design and techniques in explosive harpooning, including the introduction by Japan of a more efficient steel harpoon head which breaks into only two sections, concentrating a greater explosive force into a smaller area and killing by shock rather than by damaging major organs.

The use in the harpoon head of liquid carbon dioxide released under high pressure inside the whale, has also been suggested as an efficient and quick method of killing whales. Accuracy by the gunner would be required as the carbon dioxide is most efficient when released in the thoracic cavity of the whale where it should ensure instant death. This method has not yet been tried (IWC/SC/29/Rep. 5, 1978).

The main conclusions reached by the subcommittee were:
It would appear from available data that explosive harpooning is still the most reliable and efficient method of killing whales practised today. The use of tethered or untethered 'killer' harpoons to administer a rapid coup de grace to whales made fast, but not killed by the first harpoon is a commendable humane practice.

Other methods such as electrical harpooning and the use of drugs to kill whales should not be overlooked' (IWC/30/4, Annex J, 1978).

The eventual result of the subcommittee's deliberations was the adoption by the Commission of recommendations on humane killing. Of particular relevance among these are:
 (a) a systematic investigation and evaluation of the efficiency of present methods of killing whales is needed, in particular observations by suitably qualified veterinarians and others of

how quickly the whale loses consciousness and dies and of the nature of the injuries caused; and

(b) further research by qualified personnel into electrical, pharmacological and explosive methods should be urged on whaling nations in an effort to achieve the most humane methods of killing whales as quickly as possible.

Whaling nations have also been asked to provide further information on times of death and the reliability of the killing device used.

Humaneness

Suffering of pain, both physical and mental, is a process that mankind is all too familiar with. Similar suffering in other members of the animal kingdom is something that must be recognised, in particular for mammals. Although the Office of Regional Administration, Albany, has maintained that there is no evidence that whales are more aware of their plight, or more sensitive to pain than other animals, a British Committee on Cruelty to Wild Animals has said:

> Pain is the sensation mediated by a distinct family of nerve fibres which have their own connections with the brain....Some mammals are known to have, and all may be presumed to have, the nervous apparatus which in human beings is known to mediate the sensation of pain, and this is acceptable evidence that mammals do indeed feel pain. Further, animals squeal, struggle, and give other 'behavioural' evidence which is generally regarded as the accompaniment of painful feelings. Evidence of this second sort is, perhaps, less certain, because outward signs of pain are variable and may be absent and it is impossible to say whether, and in what sense, the cry of an animal is to be given the same weight as the cry of a human being. Nevertheless we believe that the physiological, and more particularly the anatomical, evidence fully justifies and reinforces the commonsense belief that animals feel pain (Appendix VII, Littlewood Report on Experimentation on Animals, cited in Hutton, 1978).

Many submissions also argued that whales experience mental suffering through being hunted or injured, or through the loss of companions (whales being very social animals), particularly because of their highly developed brain and nervous system (see for example Project Jonah, 1978; Hutton, 1978).

The length of time before the death of a whale, quite apart from the pain and suffering experienced before death, is one measure which can be used when examining the question of humane killing.

Best has collected data for 140 sperm whales in August to September 1971 and March to April 1973. His research indicated that for the three gunners observed, an average of 1.6 harpoons were used per whale, with 46.1 per cent of all whales being killed by one harpoon (for the small minke whale this percentage rose to 85.7 per cent); death times ranged from 'instantaneous' to a maximum of 19.5 minutes; and the median time to death ranged from 2 minutes 10 seconds to 4 minutes (depending on the gunner). While 77 per cent of whales were killed within 15 minutes, only 15 per cent were killed in under one minute. No significant differences were found in the death times for small (less than 40 feet), medium (40-45 feet) or large (over 45 feet) sperm whales (Best, 1975).

Ohsumi (1977) collected mean times to death for 244 fin whales (6.12 minutes), 564 sei whales (4.0 minutes) and 946 minke whales (3.72 minutes). Best reported an average of 5.2 minutes for times to death for 10 minke whales taken off South Africa.

> Possible reasons for this time being somewhat greater than the Japanese average is that though both use a cold (non-explosive) harpoon the Japanese employ a two-point electrical lance to kill the animal after having drawn alongside it and made fast (IWC/30/4, Annex J, 1978, p.2).

In accordance with decisions of the International Whaling Commission, the Australian Government asked Cheynes Beach to provide data on time to death and number of harpoons used for whales taken from the beginning of 1978. These show that of the 420 sperm whales taken to 9 September 1978, 195 or 46.4 per cent were killed by one harpoon; that an average of 1.7 harpoons was required per whale; and that 47.4 per cent of the whales died within one minute, 27.4 per cent within 2 to 4 minutes, 19.3 per cent within 5 to 7 minutes, 4.8 per cent within 8 to 10 minutes and 1.2 per cent in over 11 minutes. One extreme case took 15 minutes and four harpoons to die. The average time to death was about three minutes and the median about two and a half minutes. These times are shorter than those found by Best and Ohsumi.

Mr A.G. Cruickshank, skipper of the catcher Cheynes III who gave evidence at the Albany hearing (Inquiry Hearings, pp.48-92), considers that 'in the majority of cases, that is more than half of them', the whale dies instantaneously (Inquiry Hearings, p.55A). While this is more than Cheynes Beach's statistics indicate, it may be explained to some extent by Mr Cruickshank's remark that the killer harpoon is 'quite often put in to make sure' (Inquiry Hearings, p.54). It is certainly clear, both internationally and for the Cheynes Beach operation, that instantaneous death does not occur in many more than half the cases.

It is also helpful to consider the data on times to death in the light of the circumstances in which the whale is harpooned and dies. According to Mr Cruikshank's description of the chase and killing of the whale, the catcher slows to half speed when the whale is about 150 metres away, and when the whale is only about 50 metres ahead speed is reduced to dead slow. The gunner fires the harpoon at a range of about 15 metres. The catcher continues to drift forward, but the engines are then usually set to go astern. At this stage, depending of course on sea and weather conditions, the whale is normally aft of the catcher. It is then winched in to a position beneath the bow where, if the gunner has any doubt as to whether the whale is dead, a killer harpoon is fired. The time taken depends on the condition of the sea and the struggles of the whale if it is not killed outright. With everything going right for the whalers, a first killer harpoon, if needed, could be fired in three minutes but normally the process requires five to seven minutes. A minimum of two minutes must elapse because this is the time required to prepare and load the harpoon.

From the observations of the Inquiry, given a moderate sea and a skilful gunner, Project Jonah's statement that 'it would be very likely that one harpoon would kill the whale very quickly' is borne out (Project

Jonah, 1978, p.3D/14). But sea conditions vary and the practical difficulties that face the gunner, who must shoot at a moving target while mounted on a platform in a catcher very responsive to the heaving seas, are only too obvious. Many accounts have been published of the agonies of whales and long-drawn-out deaths which may have occurred in rough seas. It is appropriate to quote an account from an article published in the *Age* on Saturday, 18 June 1977, by a journalist, Mr John Larkin, in which he describes the death of two whales which he observed from one of the Cheynes Beach catchers in 1977. He refers to the harpoon being fired and striking the whale and the 'almost simultaneous explosion', and continues:

> With a great swirl of blood and foam and fear the whale dived straight down, no time for fluking, thus opening the hooks on the harpoon to gouge into it further and grip it.
>
> It kept diving and the chaser slewed around to keep the shuddering rope in front of it.
>
> Wires cracked all around the decks and winches screamed and men went everywhere. The sea was red all around us as the whale struggled to live, but died.
>
> The killer harpoon was quickly fitted into the gun in case, but there was no need, as the whale floundered just below the surface, being dragged back up. By the time it was out again they said they knew it was all over. They die, they said, when they open their mouths and close their eyes.

Larkin then describes the second kill:

> The harpoon seemed to pass right through it, which can happen, and the second explosion took longer.
>
> The whole event this time seemed in slow motion.
>
> The whale dived, and a great green cloud burst up to the surface. Blood turns green underwater at 50 feet....or was this some of its intestines?
>
> It came up on the starboard side, its huge head, a third of its total body size, shaking itself, and then it gave out a most terrible cry, half in protest, half in pain and then it dived again.
>
> They loaded the next harpoon, the killer, but could not get a shot at it as it twisted and turned, hurting itself all the more. Finally, the lookout in the crow's nest shouted down that it was coming up dying. Its mouth was opening.

They pulled it in and lashed it to the starboard side. Passing steel cable through its tail and cutting off the flukes (Project Jonah, 1978, pp.3D/18 and 19).

These passages, emotive as they are, are cited because they illustrate the fact that the use of one harpoon only does not necessarily indicate either an instantaneous death or one free from pain.

Despite what has been maintained in several submissions to the Inquiry, there is a significant difference in the methods used for killing whales and the humane practices required by law for the slaughter of cattle, sheep and pigs. In abattoirs and most slaughterhouses the animal is stunned instantaneously and then immediately killed, dying while still unconscious. The only variation from this is in a small number of country slaughterhouses where stunning equipment is not available and an alternative approved method such as a firearm is used. Detailed information on this point was provided by Mr T.J. Donoghue, a technical advisory officer to the New South Wales Meat Industry Authority with also many years experience as a Commonwealth meat inspector (Inquiry Hearings, pp.189-95).

The contrast may have arisen because the whales are killed at sea and out of sight of the public. Although man has a natural affinity with those animals which share his life on land and has devised humane methods of killing them when necessary, the public conscience has not been much exercised over the mode of killing of fish or marine mammals at sea. However, over the years and certainly throughout this century, many who have been involved in whaling have not failed to draw attention to what did appear to them to be the infliction of a most cruel death. (See for example Lillie, 1955; Hardy, 1967; Ommanney, 1971.)

It might also be argued that the methods used to ensure a humane death upon land are impracticable at sea, and no more effective method to ensure a rapid death has been devised than the use of explosive harpoons. This does not mean that the present method is acceptable, and indeed the whaling nations are themselves concerned by its inhumane aspects and are developing a specific research program on the matter.

Our conclusion then as to whether the method used to bring about the death of a whale is inhumane or not does not admit of doubt. The death of the whale is caused as a result of its organs being shattered by iron fragments from the head of the harpoon. We leave on one side the fear and terror of the chase and the exacerbation of the pain as the whale is being winched into the boat. Although death is brought about by a most horrible method, in the cases where it occurs instantaneously, the act of killing may be said to be not inhumane. But if the death is not instantaneous, or does not happen quickly, the animal is required to suffer from these truly terrible injuries for at the least three minutes and more usually up to five or seven minutes until a killer harpoon can be fired. There can be only one conclusion: that in these cases death is caused most inhumanely. The fact that these cases are a significant proportion of the total leads to the inevitable conclusion that the technique for killing whales at present used is not humane.

11. COMMUNITY ATTITUDES TO WHALING

Ethical arguments

Every civilised community lives not only by law but also by certain generally accepted moral values that apply not only to human relationships but to man's treatment of animals. In our society it is generally accepted that inflicting needless pain and suffering on animals, or killing them without due justification, is morally wrong.

Peter Singer, Professor of Philosophy at Monash University, has written extensively on ethical questions concerning human treatment of animals. In his submission to the Inquiry he sets out the position as follows: 'If a being is capable of suffering, any suffering it might experience as a result of our actions must count in our ethical deliberations irrespective of whether the being is a human or nonhuman animal' (Singer, 1978).

In considering the specific ethical arguments against whaling, Singer argues that whaling is not morally justifiable if we accept the position that 'animals should not be killed or made to suffer significant pain except when there is no other way of satisfying important human needs ...' In Singer's view the case against whaling is strengthened by the fact that the method used to kill whales is usually slow and painful and such, he contends, as would be regarded as cruel and obviously wrong if inflicted on cattle in a slaughterhouse. However, while the method presents a particularly clear objection to whaling, it is not, Singer says, the only one - so that the introduction of more humane methods would not make whaling ethically acceptable: 'In the case of a lower order of animal, we might limit our concern to seeing that the animal is killed humanely; but in the case of a member of an intelligent, social species, where the emotional links between different members of the group, and the capacity to enjoy life, are only too evident, killing is itself in need of justification, and could only be justified by some overriding necessity. In the case of whaling there is no such necessity'.

Singer's submission has been the subject of comment by Professor J.J.C. Smart, Professor of Philosophy, Research School of Social Sciences, Australian National University, and also by Dr W. Godfrey-Smith of the Department of Philosophy, School of General Studies, Australian National University. Both these philosophers refer to the inhumane method of killing used in whaling. Smart takes the view that while the infliction of such a death is indeed evil, 'it is the deprivation of further life, together with the emotional upsets to other whales, which is the main evil. That is, the killing of whales would be evil even though it could be done painlessly' (Project Jonah, 1978, p.3I/38). Godfrey-Smith generally endorses what Singer says in his submission. His conclusion, however, is that while other grounds are in fact sufficient to conclude that the killing of whales should stop, the methods now employed in the slaughter of whales constitute a very strong, and in his view the strongest, reason why that slaughter is morally indefensible and should be ended at once (Godfrey-Smith, 1978).

We received several submissions presenting a different view. In particular the submission prepared by the Office of Regional Administration, Albany, on behalf of the Albany Town Council, the Albany Shire Council, the Albany Tourist Bureau, the Albany Port Authority and the Albany Regional Development Committee, rejects what it labels the 'moral/aesthetic stance' that 'because the purported high intelligence, aural communication system and evolutionary position of cetaceans are perceived to show an affinity with man, then no individuals should be killed'. This is said to have an anthropomorphic basis, so that the killing of whales is condemned simply because they exhibit characteristics (for instance 'intelligence') associated with man, rather than a rational one. 'The only suitable criteria for weighing such moral/aesthetic/emotional reasons against scientific/economic arguments, where conflict exists, can be subjective' (Office of Regional Administration, Albany, 1978, p.14).

Similarly, the Cheynes Beach submission argues against the protectionist case that the whale is entitled to be treated as a unique species which demands special treatment because of its particular behaviour and the fact that some specific species of whale have particular behaviour patterns and brain structures indicating an ability to learn. Cheynes Beach does not regard these reasons as sufficient to put whales 'as idols' in a special class rather than, as in the case of other animals, treating them as a renewable resource. Their special features are to be seen as part of the necessary adaptation of the species to its marine environment.

Cheynes Beach also gives a salutory warning, against 'combining characteristics'. As the company submits, there is a tendency among conservationists to combine characteristics of several species of whale thus creating in the mind of the public a single animal with a multitude of characteristics, which give it a near human quality.

The Office of Regional Administration further argues that, if the 'anthropomorphic' view holds, then to be consistent those who use it as the basis for their opposition to the killing of whales should also oppose the killing of individuals of other animal groups, notably the cephalopods (a group of molluscs including octopus and squid) which 'in terms of evolutionary position, sensory and central nervous development and the capacity for complex behaviour' show the same superiority to other invertebrates that the whale exhibits in relation generally to marine mammals.

The final argument put for this submission's case is that there is no 'moral precedent' to support the contention that if whales have intelligence and unique characteristics then individual whales should not be killed by man.

At this stage it is sufficient to observe that, in our view, a judgment on whether it is morally acceptable to kill animals without justification and by inhumane methods is at the basis of standards of behaviour towards animals generally accepted in the Australian community.

Whales as a resource

Throughout his history man has generally regarded other animals as a natural resource, providing food and other necessary items such as skin and furs. In the last two decades, however, there has been a growing recognition of man's responsibility to preserve his environment, and in particular to preserve wildlife. While many species have been protected by this development, controlled exploitation has continued for others, including those kangaroo species that are commercially harvested where numbers are such as to be a nuisance to farmers, and certain species of fish and marine mammals, notably whales. Such species can be distinguished from animals such as cattle, sheep and pigs that are traditionally bred for slaughter.

The killing of these animals is traditionally also justified as a matter of necessity. This is brought out in an interesting passage by the writer, Edwin Muir. Speaking of his upbringing on a farm in the Orkney Islands, he continues:

> ...at the heart of human civilisation is the byre, the barn and the midden...When a neighbour came to stick the pig it was a ceremony as objective as the rising and setting of the sun; and though the thought never entered his mind that without that act civilisation, with its fabric of customs and ideas and faiths, could not exist - the church, the school, the council chamber, the drawing-room, the library, the city - he did it as a thing that had always been done, and done in a certain way. There was a necessity in...the killing which took away the sin, or at least, by the ritual act, transformed it into a sad, sanctioned duty (Muir, 1954, p.36).

The question confronting the Inquiry is then whether the whale is a species of wildlife that should be absolutely protected or whether it may properly be exploited as a renewable natural resource.

The Western Australian-Government believes that 'whales are a renewable resource which should be available for exploitation by man provided such exploitation is controlled', and points out in its submission that the

Government has accepted the International Whaling Commission as 'the proper authority to determine the allowable levels of exploitation' (Western Australian Government, 1978, p.4).

Those in favour of treating the whale as a protected species argue that the International Whaling Commission has failed to adopt proper conservation measures and that its present methods of estimating whale populations for conservation purposes are not to be relied upon.

Cheynes Beach, in its submission, supports the Commission, which it regards as well suited to provide a proper scientific assessment of whale stocks and quotas, and argues that whales are certainly a resource that should be harvested upon sound general farming principles whereby 'breeding stock is maintained to ensure a capability for regeneration while allowing the harvesting of the surplus'. The Company recognises the need to consider the conservation of the general marine environment and to avoid any risk of extinction of the species.

Whale products provide for certain nations a direct food resource and generally satisfy industrial needs. Cheynes Beach maintains that it is not for Australia to deny to any nation the right to harvest whales for those purposes, except of course on strict conservationist grounds and for the benefit of future generations. The submission continues that it would not be realistic for Australia to take the view that even if whales are available for harvesting they cannot be caught by other nations; the availability of substitutes is irrelevant, it is for the market place and the individual end users to determine which alternative is used; in so far as whales provide a food source it would be totally unjustifiable to suggest an alternative food source which may be outside the ability of the harvesting nation to produce for itself or which as a matter of taste it does not prefer.

Finally, Cheynes Beach argues that the consideration of whales separately from animals generally, which have always been regarded as a natural resource to be exploited by man, would raise a special category somewhere between mankind and animals as we understand them, based on an 'intellectual elitism' which is not supported by logic. Where whales do have special features these are related to the special marine environment and it is sufficient that this be taken into account in conserving numbers and the balance of the environment as we know it.

Popular attitudes towards whales

Another consideration has emerged in recent years with the growing awareness of the environment and of the need to conserve previously exploited species. It is the aesthetic appeal of whales, which has been the subject of much discussion in books and in the media, and particularly in television films. Dr R.M. Laws, the Chairman of the Scientific Consultation on Marine Mammals held in Bergen in 1976, wrote in the introduction to its Proceedings of his first encounter with whales nearly thirty years before, when he first sailed from London to the Antarctic to work on the southern seals. The passage is cited, unemotional as it is, bearing in mind that Dr Laws has expressed support for controlled exploitation. To use his own words:

> One of the memorable sights was the large number of whales surrounding our small ship as we crawled across the vastness of the Southern Ocean. A few years later, I wintered at South Georgia and saw something of the whales and whaling operations there. A little later still, during a season as a whaling inspector on a floating factory ship, came my first close encounter with a living blue whale in the pack ice - the others had mainly been dead on the flensing plan - and I experienced a great feeling of awe in the presence of the most magnificent of all animals. I should say that I have worked on the large African mammals and even they do not compare with the whales. At the beginning of this year, I was again in the Antarctic and saw just a handful of large whales. The depletion of the southern whale stocks is known to all familiar with marine mammals, but few will have experienced the change in these direct terms (FAO/ACMRR, 1978, p.7).

From the beginning of recorded history, whales have been regarded almost universally as objects of wonder. As the Project Jonah submission reminds us, they have been a favourite subject of artists and sculptors since the days of ancient Greece. Since ancient times, certain peoples have regarded dolphins, porpoises and often whales as sacred animals. Others have regarded them as animals worthy of special treatment. As we all know, seagulls and the spouting of whales have long been favourite subjects for children's drawings of the sea. The intricate songs of the humpback are regarded by many as having a strange and eerie beauty (Project Jonah, 1978, p.3B/99). In those countries where the migration of the gray whale takes it close to the shore there is no lack of whale watchers. An extensive tourist industry is flourishing off the coast of California. Whales have become the symbol of the environment for many people in Australia and throughout the western world. An affinity for whales has been developed, and many people regard them as intelligent, social animals, devoted to their young. Their appeal includes their sheer size, uniqueness and mystique, and their gentle and peaceful behaviour towards man. To use the words of Dr V.B. Scheffer, formerly of the US Fish and Wildlife Service and former Chairman of the Marine Mammal Commission:

> If I understand what men and women are saying today about whales it is 'Let them be'. A useful whale, they say, is one out there somewhere in the wild - free, alive, hidden, breathing, perpetuating its ancient bloodline (Scheffer, 1978, p.4).

If has been widely argued that the benefits to society of tourism, entertainment and other so-called low-consumptive uses of whales, together with their broader aesthetic and cultural value, should be taken into account in whaling management (See for example FAO/ACMRR, 1978; Larkin, 1977; Scarff, 1977.) At the recent meeting of the International Whaling Commission in London, it was suggested that the Commission should take these matters into account. At that time, the mandate of the Technical Committee was extended so that the Commissioners could choose to have the benefit of advice from a wider range of experts, including for example sociologists, economists and consumer representatives.

This report would not be complete without reference to various strong expressions of views held by the public. Over 37,000 representations calling for an end to whaling in Australia have been received by the Prime Minister during 1977 and 1978. Of these over 14,000 were from within Australia with the remainder coming from overseas (predominantly from America). Seventy-eight petitions have also been presented to

Parliament calling for an end to whaling, supporting a ten-year moratorium and calling for the closure of Cheynes Beach. This includes one from Project Jonah with over 145,000 signatures. In addition, during the course of the Inquiry, the Australian Conservation Foundation prepared a petition: 'I, the undersigned, join with the Australian Conservation Foundation to petition the Australian Whaling Commission and the Governments of Japan and the Soviet Union and all other countries engaged in commercial whale killing to put an immediate stop to the needless slaughter of whales.' It contained over 40,000 signatures.

Public opinion polls have also been conducted by Project Jonah, Cheynes Beach, the <u>Albany Advertiser</u> newspaper and the Australian Conservation Foundation. The first public opinion poll was conducted on behalf of Project Jonah by Australian National Opinion Polls in December 1977, eight days before the Federal Election. Two questions were added to the end of a regular pre-election poll. The following presents the results of this poll in tabular form:

Q1. How interested are you in the debate that is going in Australia at the moment about whether we should or should not be killing whales? Are you interested a lot, a little or not much?

	Total %	NSW %	Vic %	Qld %	SA %	WA %	Tas %
Interested a lot	37	37	41	34	41	24	35
Interested a little	34	34	31	35	34	43	35
Not much	27	28	25	30	23	32	28
Unsure	2	1	3	1	2	1	2

Q2. Do you think Australia should or should not give up the killing of whales now?

	Total %	NSW %	Vic %	Qld %	SA %	WA %	Tas %
Should give up	66	67	75	61	76	39	53
Should not give up	19	22	12	15	11	41	23
Unsure	15	11	13	23	13	20	24

The poll shows that 66 per cent of all Australian voters thought Australia should give up the killing of whales now. In every State except

Western Australia, in which the whaling station was situated, there was a clear majority in favour of giving up the killing of whales. In Victoria the figure was as high as 75 per cent, and in South Australia 76 per cent. It is relevant to note the high degree of interest shown in the question whether whales should be killed or not. This may have been due to the widespread public interest in the whole question of Australian whaling, and whaling generally, following the meeting of the International Whaling Commission in Canberra in June 1977. Newspapers and magazines had featured the views of both the protectionists and the whaling interests. Large advertisements had been inserted in major newspapers under the headline 'The Australian case for whaling'.

The Cheynes Beach survey was conducted by Australian Marketing Services Pty Ltd, and covered Perth, Albany and Melbourne. One thousand respondents were selected at random, with 400 each in Melbourne and Perth and 200 in Albany. Interviewing was conducted in February 1978 in Western Australia and April-May 1978 in Melbourne.

The following table presents a summary of overall attitudes to whaling:

Question: In an overall way, which of these statements best sums up your attitude to whaling in Australia?

	Melbourne %	Perth %	Albany %
Whaling should continue - I entirely approve	4.5	10.9	50.0
I approve - but have some reservations about it	19.7	39.1	33.5
I am undecided	14.7	14.9	7.5
I disapprove, I don't think it is a good thing	31.2	16.4	7.5
I strongly disapprove	29.9	18.7	1.5
	100.0	100.0	100.0
Base (number):	(401)	(402)	(200)

It will be noted that there was strong support in Albany for the continuance of whaling, and somewhat less in Perth. So far as those interviewed in Melbourne are concerned, there was a clear majority who disapproved of the continuance of whaling.

The <u>Albany Advertiser</u> in April-May 1978 included coupons inviting readers to express their views on whaling. A total of 271 coupons were returned of which 185 were in favour of whaling continuing and 75 against, with 11 informal. This represented 68 per cent in favour of whaling continuing in Albany. The questions asked, with a simple yes/no choice in each case, were:
- (a) 'I think that the killing of sperm whales by the Cheynes Beach Whaling Company should cease and that the taking of sperm whales off Albany is unnecessary' and
- (b) 'I think that the Cheynes Beach Whaling Company should be allowed to take whales under present controlled conditions, due to the economic importance of whaling to Albany'.

While these coupons do not necessarily provide a representative sample of opinion, the strength of support shown within Albany for whaling is similar to that found in the Cheynes Beach survey. It must of course be recognised that many Albany people have a financial interest in the continuation of whaling.

The next opinion poll was taken in June and July 1978 on behalf of the Australian Conservation Foundation by national poll consultants, Irving Saulwick and Associates. It covered 2923 respondents in all federal electorates, except the Northern Territory, proportionate to the numbers of electors in each electorate. Dr J.G. Mosley, the Executive Director of the Australian Conservation Foundation explained that the purpose of this poll was to cover questions which the Foundation believed were pertinent to the Inquiry's terms of reference. In particular the Foundation was concerned that without further information it could be argued that none, or an insignificant proportion, of the 66 per cent of Australians revealed by the Project Jonah as opposed to whaling, would be opposed to it if it could be shown that changes in the method of regulating the kill would keep the species from being threatened. In other words without further information the possibility would always remain that the main reason for opposing whaling was concern about the threat to the species and the risks involved in present management procedures. The poll in fact showed that about 42 per cent of Australians believed that whales should not be killed at all, even if it could be shown that whaling does not threaten the existence of the species (Inquiry Hearings, pp.779-80). The results of the poll are presented in summary in Appendix 17.

As with many opinion polls qualifications have to be made to take account of the respondents' varying degree of interest in the questions asked and also their varying knowledge of the subject. However, the conclusion certainly can be drawn from these polls that there is a significant proportion of the Australian public which is opposed to the killing of whales for whatever reason.

It is not only opinion polls that show a significant support for an end to whaling. Project Jonah points to the very strong support from children and their teachers on this issue, an emotive issue as it may be for young children (Project Jonah, 1978, p.3I/8). There is also a strong group of conservation organisations, all of them with the support of very large numbers of members, arguing that whales should be protected. They include Project Jonah, the Australian Conservation Foundation, Friends of the Earth, and Greenpeace Australia. The same view is taken by the Governments of New South Wales and South Australia at least.

It was submitted by Project Jonah that on the basis of demonstrated public attitudes and opinions, the killing of whales is wrong in the eyes of the Australian community. The submission then went on to suggest that the continuation of whaling by Australia would outrage a significant proportion of the population. The latter view is one with which the Inquiry agrees.

The opposition to whaling in Australia is in line with the attitudes in many other countries. The United Kingdom, Canada and New Zealand all withdrew from whaling for economic reasons, but since March 1973 the United Kingdom has maintained an import ban on all whale products with the exception of sperm whale oil, spermaceti wax and ambergris and New

Zealand prohibited the import of whales and whale products in August 1975. Since 1972 the United States of America has stopped whaling and has been opposed to the continuance of whaling. This policy has been strictly enforced under a number of statutes beginning with the Endangered Species Conservation Act 1969 and the Marine Mammal Protection Act 1972.

The strong body of international opinion against whaling is well summed up in the following passage by Sir Peter Scott, Chairman of the World Wildlife Fund (World Wildlife Fund, 1978, p.3).

> The UN Stockholm Conference in 1972 called for a 10 year moratorium on all commercial whaling to allow the whale stocks to recover. This is still the view of the World Wildlife Fund, International Union for Conservation of Nature and, I believe, all organisations with a basic philosophy of conservation. WWF/IUCN launched a worldwide campaign 'The Seas Must Live' in an effort to stimulate Governments into taking action on a number of marine conservation issues (of which whales and whaling was one - and an especially important one), before it was too late. Even in developing countries today's young people have acquired a surprisingly enlightened attitude towards conservation. They seem in many parts of the world to have realised that conservation is absolutely <u>necessary</u> if the options are to be kept open for their children and grandchildren. In this the whales are a powerful talisman. As one sixteen year old said to me: 'Although I've never seen one, I want to make sure that the great whales are still around for my grandchildren to see and haven't been exterminated by you lot, who are supposed to be our elders and betters.'

12. AUSTRALIA'S FUTURE POLICY ON WHALING

It remains to consider the Inquiry's conclusions on the various arguments for and against whaling - aptly referred to by Cheynes Beach in its submission as the 'whaling debate' - and to express our final opinion on Australia's future whaling policies.

In the course of the Inquiry two issues have emerged as dominant in shaping our conclusions. These are the severely depleted state of the world's whale stocks, which has been brought about by their continued harvesting without adequate regard to the risks inherent in the exploitation procedures followed, and the special nature of the whale itself including the real possibility that we are dealing with a creature which has a remarkably developed brain and a high degree of intelligence.

It is against the background of these two considerations, then, that we have assessed the arguments put before us about the proper future policy to be adopted towards whales.

The main arguments relied on by those in favour of the conservation of whales have all been covered in previous chapters. They are the special features of whales and their potential for intelligence; current community values which have led many people to oppose whaling; the risks shown to exist in the management procedures of the International Whaling Commission; the existence of substitutes for whale products generally and of the necessary technology to replace sperm oil in the few remaining applications where it is still used; and the inhumane methods used to kill whales.

Arguments opposing the case of the conservationists have also been reviewed. Prominent among these are that whales are no more special than some other animals and therefore can be considered a renewable natural resource; the market should decide whether whale products are used or not;

under present management procedures whales are not threatened with
extinction and can support a continuing harvest; communities using whales for
food should not be denied this; and the methods used to kill whales are
acceptable. In addition, the Inquiry's attention was drawn to the drastic
curtailment of the Japanese industry. Three whaling fleets have been reduced
to one, and about 10,000 people who had been directly engaged in whaling have
been forced out of the industry (Japan Whaling Association, 1978; Nishiwaki,
1978).

Finally, the argument was presented in the submission of the Office of
Regional Administration, Albany, that the views of the people of Albany in
favour of continuing whaling should be given special weight in deciding the
future of Australian whaling, since they, of all Australians, understand
whaling and are familiar with the whale as an animal.

On some of these matters where there is a divergence of opinion we
have reached conclusions in earlier chapters; others we have deferred
until now so that the individual issues can be considered in a wider context.

World whale stocks

The initial step in reaching our conclusions has been to look at the state
of the worldwide whale stocks. The whaling industry depends on the
availability not of whales generally, but of a limited number of stocks.
We have already noted the state of the stocks (Chapter 6 and Appendix 10)
and the general picture is that they have been excessively exploited.
Several of what were large stocks of whales have been protected for many
years and now survive in very low numbers with no signs of substantial
recovery. They are the blue whale - the largest animal that has ever
lived - and the humpback, right and bowhead whales. Of those stocks which
have been long protected only the California gray whale has shown a
substantial recovery. It is a simple matter for the International Whaling
Commission to leave these stocks on one side now they are protected, but we
have seen how devastating the reduction was - from about 450,000 whales of
exploitable size to less than one-tenth of this, with the largest stocks
being reduced by more than 95 per cent.

Whaling of fin whales in the southern hemisphere was ended only at the close of the 1975-76 season. These whales are now protected in all oceans except in part of the north Atlantic. Sei whale stocks in the southern hemisphere had by 1977 been reduced to levels where protection was necessary for four of the six Areas, and the stocks in the two remaining Areas were added at the 1978 meeting of the International Whaling Commission, leaving a small stock adjacent to Iceland as the only sei stock subject to exploitation. The reduction in fin and sei stocks has also been substantial - from about 600,000 whales of exploitable size to about 150,000. Of the great whales, only the sperm whale is still widely found in large numbers. Female sperm whale stocks have in general been exploited lightly but males have been extensively exploited and in several cases reduced to levels that cause concern to scientists.

The only other stocks available for harvesting are the Bryde's and the small minke whale. Some stocks of Bryde's whales may have been reduced to some extent by land-station operations but these whales are lightly exploited in general. Minke whale stocks seem to be expanding in some areas and maintained at present levels in others by exploitation. In the days when the blue, the fin and the sei whales were hunted, the minke was regarded as a 'dwarf' and the whaling industry showed no interest in it (Slijper, 1962). So the whaling industry is really scraping the bottom of the barrel, and considering its drastic curtailment it is now a 'dying industry' (Monitor, 1978). The extent of this curtailment is shown by a comparison of present day catches for the southern hemisphere with those of a decade or two earlier. For the 1957-58 season and the three following an average of about 27,700 fin whales were taken in the Antarctic, and for the 1964-65 season and the two following an average of about 16,800 sei whales were taken there. Southern hemisphere sperm whale catches averaged over 12,000 per year for the two decades to 1974-75. Now the whaling industry is left only with the present quotas of about 5400 sperm whales and about 6200 of the small minke whales in the southern hemisphere. But it remains undeterred by the fact that it is the exploitation by modern methods only over the last half century or so which has caused the huge reduction in the stocks of whales.

Management of whales

No doubt from the viewpoint of the whaling nations, proper management of whale stocks means that exploitation should be limited to the sperm, minke and Bryde's stocks until the fin and sei whale stocks recover from their present protection status and are considered to be ready for another period of exploitation. For most stocks, if the Commission's assumptions that recovery can occur are borne out, this does not seem likely to happen for some years (Chapter 6). But it is the overall picture which is significant. While the present management procedure largely eliminates the risk of actual extinction of any stock, there are still many aspects of the scientific advice and modelling procedures in which there is uncertainty and scope for error (see Chapter 4). This applies at even the most basic levels. Sir Peter Scott, in his statement on behalf of the World Wildlife Fund to the 1978 meeting of the International Whaling Commission, singled out the critical population level below which a stock could not survive as one matter on which there was a 'profound ignorance'. Although the polygynous nature of the sperm whale may make it a special case, there would seem to be little room for error when, as is the case for most sperm whale stocks, it is planned to reduce the number of mature males to 32 per cent of its initial number. (In fact as we have seen, in the Division 5 stock the number has been reduced to 26 per cent.)

A major element of risk remaining in the Commission's procedures cannot be ignored. It arises from the use of single species management when the whole ecosystem should be taken into account. Ecosystem considerations are of special significance, as we have seen in Chapter 4, for the recovery of the long protected baleen stocks including the humpback and right whales which for the early settlers were a familiar sight off Australian shores and in Australian harbours.

We have noted the plans for a major fishery upon the Antarctic krill on which these whales depend for food. The depleted whale numbers have also enabled other species, such as the crabeater and fur seals and the penguins, competing for the same food supply, to gain an advantage which may prove to be permanent. We have noted Tranter's view that the smaller animals may have the edge in that competition because of their higher capacity for increase, and that there is no certainty in the Commission's assumption that when whaling ceases on a stock the numbers will eventually increase (Chapter 4). So the capacity of the remnants of the blue, right and humpback whales to recover to anything like their original numbers is in real doubt, and in the competition for the krill there is a risk that stocks which have been reduced to low levels may decline further. Man cannot intervene in the marine environment as he can with land animals and establish physical conditions which would ensure conservation. An attempt could be made to manipulate the ecosystem throughout the Antarctic so as to hold the competitive species in check, but the technology and the economics of it are doubtful.

The sperm, minke and Bryde's whales have suffered the least from exploitation. The fact that the minke whale seems to have been increasing in numbers before exploitation, as we see the situation, should not be taken as further justification for its exploitation. In the interests of ensuring that the whale in at least these limited species should continue to be abundant for future generations of mankind, it is our opinion that the time has come to call for a halt to whaling. In 1970 acting on a similar concern for the whales' continuance, the United States Government considered it necessary to place eight species of whales upon the list of endangered species under its endangered species legislation.[1]

The Inquiry has found support for its view in the sound warning given by Mr D.J. Tranter, Senior Principal Research Scientist, CSIRO. To use his own words 'Until such time as research is adequate to define more closely the limits of certainty in managing the exploitation of multi-species resources, the appropriate form of management to maximise our options needs to be conservative. To gamble with uncertainty is defensible only when our demand on the resource is paramount and urgent. Is this the case with the baleen whales?' As he says, 'If greater certainty is desired there is a ready option - stop the harvest' (Tranter, 1978, pp.2-3).

---

(1) We note in the last stages of this Inquiry that the Soviet Union has cut its Antarctic whaling fleet for this season to help restore whale stocks (Tass news agency, Canberra Times, 3 November 1978).

The threat to the baleen whales is undeniably immediate. New pressures - after millions of years of evolution - face the whale and they are likely to increase. Substitutes for whale products now exist and man can afford caution. Subsistence or aboriginal whaling is a special case. The catches are small and the cultural and social benefits may be sufficiently important to vary general management practices. So far as commercial whaling is concerned, the Inquiry considers that a call for its end is amply justified.

Such a call is, in the Inquiry's opinion, the best way in which Australia might pursue its policy of preservation and conservation of the many species of whales. As will be seen from an examination of the other arguments placed before the Inquiry, such a policy is also required by the general community values of the Australian people.

The principal arguments for and against whaling

Man's hunting of wild beasts and his animal husbandry has traditionally been justified as providing necessities such as food and clothing. This is entirely consistent with Australian attitudes to whaling in 1949 - a year, as we have seen, of some significance for Australian whaling. The first whaling Acts were already on the statute book. It was then that, as already noted, the Australian Parliament authorised the establishment of the Australian Whaling Commission to carry on a whaling industry upon the shores of Western Australia. A perusal of Hansard shows that the reasons for the proposal were, naturally, the economic advantages this expansion of primary industry would secure, and also the supply of the world market for whale products which, in the absence of substitutes, were essential for some industrial purposes. The measure had the support of all parties, and no suggestion was made that 'the whaling industry should not be exploited' (Commonwealth Parliamentary Debates (H of R) Volume 203, 1830-45, especially at 1842). But, going behind the legislation, its basis could only have been that the Australian people were content at that time to regard whales as a natural resource, and apply to whales a view which has been expressed by some parties during the course of the Inquiry, that 'mankind has always taken a crop from animals'. It can also be assumed that no objection was then taken to the techniques of killing whales. So that venture, which was the beginning of the modern Australian whaling industry, involved Australia as a whole.

But as we have noted in this report (Chapter 11) there has since those years been a widespread change in man's attitude to the environment and to wildlife, and a renaissance in the popular mind of the conception of the whale as a gentle peaceful creature, intelligent and friendly to man which was the view of older civilisations.

The maintenance of our natural heritage to be passed on to subsequent generations is now regarded as a priority. In most countries where recognition has been given to these changed values and where policies opposed to whaling have been adopted, circumstances enabled the decision to be made without the compulsory closure of the local whaling enterprise with consequent hardship to employees. We note, however, that this was not the case in the United States (Chapter 8). The Australian people also have developed new attitudes to the animals with which we share our world environment and these are now embodied in the statutes enacted by the Commonwealth and the States for the protection of all wildlife (Chapter 3). With the closure of Cheynes Beach, unfortunate as it was for those dependent on the industry, in fact the path is made easier for Australian policies on whaling to be changed in accordance with those attitudes.

When, as in this case, the task is to make policy recommendations, what are the criteria to be applied? The decision must surely be one which would commend itself to reasonable and responsible men and women of Australia, and must therefore be reached with due consideration of prevailing community values. But these values should be such as are generally accepted in Australia as a nation, and not values confined to any one region.

The main issue is whether whales should be regarded as a renewable resource available for exploitation by man, subject of course to proper control - an issue which was outlined in Chapter 11. Those who argue in support of whaling do not accept that the availability of substitutes for whale products is any answer. It is the satisfaction of a human need, in this case the use of whale products for certain valued processes, which is put forward as a legitimate justification for whaling.

In approaching this issue consideration must be given to the kind of animal the whale is. In practice, man does draw a distinction between animals. A century ago Darwin used rather a broad classification between 'the higher animals' by which he meant the higher mammals, and 'lower animals' (Darwin, 1883, Chapters 3 and 4). For present purposes it is possible to list some wild animals which do have a special significance for man - and the list is of course not exhaustive - including the elephant, the primates, the lion, the tiger and the koala. Some of these animals derive their significance because they are noted for their mental powers, some for their grace and beauty, and some for their special appeal. It is not necessary for the animal to demonstrate 'intelligence' to be particularly valued. This practice may be elitist but it is grounded in man's instinctive attitudes and our pragmatic approach demands that we take account of it. If exploitation of animal resources were to proceed uniformly and without the recognition of any such special classes, the elephant could again be hunted for its tusks, and the koala, protected in Australia now for 50 years, for its skin. Yet in either case the renewed hunt would draw worldwide condemnation.

It is not necessary for our task to indicate where the line is to be drawn. The choice in relation to any animal can be left for man to make when the situation arises. But when we turn to the whale - and we include dolphins and porpoises - there can be no doubt that it is one of those animals which for man have a special significance. We have noted that the whale has been evolving for at least 25 million years - a considerably longer period of time than man. Separated in its marine environment, quite remarkably, it has developed a brain with structures so similar to that of man that a real possibility must be taken to exist that the whale has the potential of high intelligence. The whale's behaviour is also consistent with a highly developed brain, but again we must await further evidence, especially on the meaning of the sounds by which whales communicate. In the meantime the only reasonable course is to make allowance for the fact that we are indeed dealing with special creatures. From a biological point of view, the whale is one of the 'two mountain peaks of evolution on planet earth - on land,... human beings and in the sea, cetacea' (Connecticut Cetacean Society, 1978, paraphrasing Dr Teizo Ogawa of Tokyo University).

The pragmatic approach outlined above has perhaps foreshadowed our conclusion; it is supported by the prevailing community values to which we shall now turn.

We are unable to agree with the view that the propriety of killing whales is a matter merely of individual and subjective judgment which has no place in the formulation of policy. There is a wide category of animals which the community considers it is wrong to kill. The argument that 'it is morally wrong to kill whales' was adverted to by Mr J.L. McHugh, a former United States Commissioner to the International Whaling Commission. He remarked that he could not quarrel with this view on philosophical grounds. His answer was, 'The most telling argument against it is that the concept is too idealistic for an imperfect world' (McHugh, 1976, p.407). But it seems that he underestimated the force of the worldwide concern for the whale.

Professor Singer expressed the view that animals should not be killed or made to suffer significant pain except when there is no other way of satisfying important human needs (Chapter 11). His view does not, in our opinion, go beyond the values commonly accepted by the Australian people in relation to animals such as whales. (Of course, it is entirely in keeping with this proposition that there should be certain exceptions for example, in the case of people acting in self defence, in the interests of the animal itself or of the animal population if the numbers need to be culled or, as mentioned in Chapter 11 in relation to kangaroos, if the species is encroaching upon what are regarded as man's legitimate interests.)

There has been a further change since the starting point of modern Australian whaling in 1949. In the 1970s we have seen adequate substitutes for whale products, or the technology to supply them, developed. The Inquiry considers that in the light of work already done in this area, a reasonable period for the development and testing of substitutes for use in all circumstances where whale products are currently used in Australia would be two years. While the position would be similar in other industrial countries, there may be economic or technological reasons which would prevent some countries from switching quickly to substitutes, and also some communities would wish to continue to take whales in small numbers for social and cultural reasons. We have not overlooked the fact that whale meat plays a part in the diet of some nations, and we sympathise with the adjustments which would be required if they were to stop using whale products. However the amount of protein contributed to their diet by whale meat is generally only a relatively small percentage of the total, and, given the availability of alternative foods, we would be hopeful that, in the long term, this factor would not be seen as supporting the continuance of whaling.

Upon all the evidence, then, it cannot now be said that whaling is justified as the only means of satisfying important human needs. Australia's present community attitudes therefore dictate that the whale should not now be regarded as a natural resource available for exploitation. We are confident that, in the light of all the facts put to the Inquiry, reasonable Australian citizens would conclude that, now there is no necessity, it is wrong to kill an animal of such special significance as the whale.

An additional argument is to be found in the fact that, while significant proportions of the Australian people support both the case for and the case against whaling, it appears that the majority when questioned directly on the issue are against whaling. We support the view of Scheffer (1978), who suggested that it is for the wildlife manager to listen carefully for the majority voice, to anticipate trends in public opinion and to recommend what he considers the appropriate action.

There is another argument which was mentioned in the news media after the closure of Cheynes Beach. It is also the argument which, it seems, Nishiwaki had in mind when he referred to the heavy loss of employment caused in Japan by the reduced quotas. It is that 'people are more important than whales'. This argument sets aside all other considerations and maintains that there is a higher and overriding community value, grounded on the welfare of a man and his family, that jobs come first. The argument would have required earnest consideration, particularly in these times of economic difficulty, had Cheynes Beach not reached its own independent decision to close for commercial reasons. But for the Australian whaling industry at the present time, it is really based on expediency. It does not now arise for consideration.

Another concern raised was that if Australia did not take the whales off Albany other countries would. In the meantime, of course, there has been the Australian analysis of the stock as justifying immediate protection (Chapter 6). The whales would also be afforded a continuing measure of protection by a prohibition of whaling within the 200-mile fishing zone, an area which certainly includes the Cheynes Beach operations.

Several other arguments were put forward against bringing whaling to an end, but these did not weigh against the main thrust of our conclusions. There is every reason to suppose that in the International Whaling Commission, Australia will play the same effective role as the other non-whaling countries, and the loss of data for research which is provided in the course of the hunt and from the slaughtered whales is not a consideration which can outweigh the matters of principle upon which the Inquiry's views are based.

Finally, our summing up would not be complete without a reiteration of our earlier conclusion that the time has come to recognise that the method used to kill whales is inhumane. This is another area in which community attitudes have changed since 1949, when Parliament put itself behind the whaling industry in Australia. No doubt in those days the standards prevailing in slaughterhouses fell short of those of the present day. However, as noted in Chapter 10 the humane methods now required by law in slaughterhouses ensure that the beast is stunned and unconscious when killed. Not even approximately the same standards can in practice be obtained at sea when killing whales, and for this reason we believe that the killing should stop. We should emphasise, none the less, that even if these problems were overcome completely, the Inquiry would still consider on the other grounds already set out that a protection policy was proper.

This Inquiry has then reached the conclusion that Australia's policy should be changed, and that Australia should oppose the continuance of whaling, both within Australia and also abroad. Whales are migratory animals and are the heritage of all nations. We have a national and an international responsibility to preserve them for future generations.

CONCLUSIONS AND RECOMMENDATIONS

The Inquiry's central conclusion is that Australian whaling should end, and that, internationally, Australia should pursue a policy of opposition to whaling.

An immediate issue concerning the Division 5 sperm whale stock, which has been taken by Cheynes Beach off Western Australia, is the position Australia should adopt at the meeting of the International Whaling Commission to be held on 19 and 20 December 1978. Catch limits for the 1978-79/1979 season were set for this Division by the Commission in June 1978, but a more recent assessment by Australian scientists shows that males in this stock should be classified as a Protection Stock, and that it is highly desirable that females also be given immediate protection. Australia should urge the Commission to take account of this new evidence and classify both males and females as Protection Stocks and amend the catch limits for this Division to zero.

A similar scientific concern may arise at the same Commission meeting about the state of the north Pacific sperm whale stocks. Australia should be prepared to press for caution in reviewing catch limits for these stocks also.

In accordance with the Inquiry's central conclusion, whaling should be prohibited within Australia's 200-mile fishing zone. Assuming that at some time the Division 5 stock recovers and is exploitable under international management procedures, this would provide the stock with some measure of protection, as it would do for the Division 6 stock in waters off the eastern States of Australia. This policy would also provide further support for the protection given to humpback and right whales through the Commission.

If the waters off the Australian Antarctic Territory are included in the fishing zone, a policy of prohibition of whaling should also apply in these waters, although we recognise that broader international considerations may affect its immediate implementation.

More generally, Australia should seek to achieve greater worldwide protection for all cetacea. The taking or killing of any cetacea - whether intentionally for scientific, display or other purposes, or incidentally such as in fishing or shark netting operations - should be carefully scrutinised to ensure that it is either essential or unavoidable. The Commonwealth should seek the cooperation of all State Governments in implementing this policy within Australia, especially in regard to small cetacea.

While these objectives could be met by amendment of the Whaling Act 1960, consideration should be given to replacing this Act by legislation specifically directed to the protection of cetacea. It may be desirable to extend such legislation to cover all marine mammals, as has been done by some other countries, but this is beyond the terms of reference of the Inquiry.

Consistently with the new policy, and to reduce the incentive for other countries to continue whaling, Australia should ban the import of whale products, as has already been done by the United States and New Zealand, and to a substantial extent by the United Kingdom. This could be done under the Customs (Prohibited Imports) Regulations. Implementation of the ban should however be deferred for two years to enable users of sperm oil and spermaceti to revise existing formulations and change over to substitutes.

We have dealt in Chapter 8 with the question of compensation for the closure of the Cheynes Beach operation. The closure, we appreciate, is of great concern to the Albany region and particularly to the Company's employees. We would recommend, and indeed the Government has already so offered, that the Commonwealth Government cooperate with State and local authorities to provide assistance under the range of programs it has available. However, the closure was inevitable if only because the whale stock was reduced to a level requiring protection, and accordingly the Inquiry is unable to recommend compensation.

Internationally, Australia should join with the many nations opposed to whaling and use its influence to encourage other countries to stop whaling. In particular, Australia should remain a member of the International Whaling Commission which is the forum where its anti-whaling policy can best be pursued. Chapter 5 reviewed the history of international management of whales, and we noted there the general support for Australia's continued membership of the Commission.

However, consideration must be given to special circumstances in some countries such as the need for whale products for food. A long-term exception may need to be made for subsistence catches such as those in the Arctic regions and in the south Pacific off Tonga. As most stocks on which subsistence whaling continues are at critically low levels, subsistence catches, small as they are, should be strictly controlled.

Australia should also promote greater international acceptance of the need to consider the links between whales and other elements of marine ecosystems. In particular, any proposal to exploit marine resources such as krill should be carefully examined for possible harmful implications for whales.

One major proposal put to the Inquiry was the establishment of a marine research station at Albany. This was supported in a number of submissions, including a detailed one by the Great Southern Regional Development Committee. While Albany has both advantages and disadvantages as a location for continuing cetacean research, for the reasons set out in Chapter 6, we consider it is impractical to retain Cheynes Beach's plant and catchers for research on whales. The question of facilities for other marine research, especially relating to fisheries, involves much wider considerations outside the terms of reference of this Inquiry. Accordingly we make no recommendation on this matter.

Australia should certainly play its part as a country prepared to provide the facilities for further study of species whose conservation is of international interest. A wide-ranging research program could usefully be developed in Australia to look at many aspects of cetacean biology and behaviour, and we suggest that governments and interested organisations seek to promote and support such a program. However, in the present economic climate we must take a realistic view in making recommendations which require Commonwealth Government funding.

Finally, we would emphasise that if the International Whaling Commission is to be effective it must have adequate information and expertise. Australia should continue to make a substantial contribution here.

Recommendations

1. Australia should oppose the continuance of whaling. While the Whaling Act 1960 remains in its present form, any application for a whaling licence should be refused, with the exception of a licence to take, in appropriate circumstances, a limited number of dolphins live for display purposes, and only in special circumstances should a permit be issued to take any cetacea for scientific purposes.

2. Whaling by other nations should be prohibited within the Australian 200-mile fishing zone. If Australia decides to include waters off the Australian Antarctic Territory in this zone, a policy of prohibition of whaling should also be pursued in these waters.

3. At the International Whaling Commission meeting to be held in December 1978, Australia should seek to classify both male and female sperm whales in Division 5 as Protection Stocks and to amend the catch limits presently set for this Division to zero. Furthermore, Australia should press for caution in the setting of other catch limits at that meeting.

4. Consideration should be given to the repeal of the Whaling Act 1960 and its replacement by new legislation directed to the protection of cetacea, and if thought desirable other marine mammals, along the lines of the marine mammal protection legislation of the United States and New Zealand.

5. The import of whale products or goods containing whale products should be banned in Australia from 1 January 1981.

6. Australia should continue to be a member of and support the International Whaling Commission as the most appropriate body to be responsible for conservation of whales internationally. Australia should support current efforts to revise the International Convention for the Regulation of Whaling 1946. In particular Australia should seek to extend the Commission's charter to the conservation of all cetacea. It should also support increased liaison and cooperation between the Commission and other international bodies with responsibilities affecting cetacea.

7. Australia should seek to achieve a worldwide ban on whaling. In working towards this Australia should propose a more cautious approach in setting catch limits and a greater emphasis on the conservation of whales by the International Whaling Commission. Although seeking to bring an end to whaling, Australia should take into account any adjustment required by the special needs of particular countries, such as the requirements of some nations of whale meat for protein, and subsistence catches by some local communities.

8. Both in its own planning and in international discussions on the management of marine resources Australia should ensure that any implications for whales are given consideration. In particular, any proposals to exploit krill or other Antarctic marine resources should be developed with consideration of the whole ecosystem and examined to see that the potential recovery of depleted baleen whale stocks is not prejudiced.

9. The Commonwealth Government should promote research on whales by Australian scientists and research institutions. It should provide funds at no less than present levels for this work. Priority should be given to:
   (a) the monitoring of the sperm, humpback and right whale stocks off Australia, including their abundance and matters affecting their recovery in numbers and their success in breeding;
   (b) the refinement of techniques for whale stock assessment and ecosystem modelling, especially while any whaling continues;

(c) the monitoring of Australian strandings of cetacea and the preservation and study of the specimens thus provided.

10. The continued Australian involvement in the International Whaling Commission should emphasise particularly participation in the Scientific Committee so that results of Australian research are promptly incorporated in that Committee's deliberations, and to ensure that due scientific attention is given to the various criticisms of current assessment procedures.

APPENDIX 1: PEOPLE AND ORGANISATIONS WHO PROVIDED WRITTEN
EVIDENCE TO THE INQUIRY AS SUBMISSIONS OR THROUGH CORRESPONDENCE

Organisations

Agriculture, Western Australian Department of	Western Australia
Alaska Center for the Environment	Alaska, USA
Albany Advertiser	Western Australia
Albany Chamber of Commerce Inc.	Western Australia
Albany Conservation Society	Western Australia
American Cetacean Society	California, USA
Animal Liberation	New South Wales
Animal Welfare Institute	Washington DC, USA
Arthur C. Trask Corporation	Illinois, USA
Australian Catholic Study Circle for Animal Welfare	Victoria
Australian Conservation Foundation	Victoria
Australian Greek Welfare Society	Victoria
Australian Institute of Petroleum Ltd	Victoria
Australian Littoral Society	Queensland
Australian Mammal Society	Victoria
Australian Society for the Study of Animal Behaviour	Australian Capital Territory
Berol Limited	London, United Kingdom
Bevaloid Australia Pty Ltd	New South Wales
British Leather Manufacturers' Research Association	Surrey, United Kingdom
Caltex Oil (Australia) Pty Ltd	New South Wales

Cheynes Beach Holdings Ltd	Western Australia
Coates Brothers Australia Pty Ltd	New South Wales
Commonwealth Serum Laboratories	Victoria
Connecticut Cetacean Society	Wethersfield, USA
Cousteau Society	New York, USA
Croda Chemicals Group Pty Ltd	Victoria
CSIRO	Victoria
Diamond Shamrock Corporation	New Jersey, USA
Ecology Action	New South Wales
Esso Chemicals Limited	London, United Kingdom
Federated Tanners Association of Australia	Australian Capital Territory
Frenchman Bay Whaling Museum	Western Australia
Friends of the Earth, New South Wales	New South Wales
Friends of the Earth, New Zealand	Auckland, New Zealand
Friends of the Earth, Victoria	Victoria
Friends of the Earth, Western Australia	Western Australia
Great Southern Regional Administrator	Western Australia
Hamilchem Pty Ltd	Victoria
Harrison Manufacturing Co. Pty Ltd	Victoria
Harrisons and Crosfield (Aust) Ltd	Western Australia
Henkel Australia Pty Ltd	New South Wales
Highgate and Job Ltd	Paisley, Scotland
ICI Australia Limited	Victoria
International Society for the Protection of Animals	London, United Kingdom
International Union for Conservation of Nature and Natural Resources	Switzerland
Japan Whaling Association	Tokyo, Japan

Jojoba Plantations	Victoria
Keil Chemical Division, Ferro Corporation	Indiana, USA
Kiwi Australia Limited	Victoria
Lubrizol Australia	New South Wales
Maritime Workers' Union of Western Australia	Western Australia
Mayco Oil and Chemical Company, Inc.	Pennsylvania, USA
Mobil Oil Australia Limited	Victoria
Molybond Laboratories	Victoria
Monitor	Washington DC, USA
National Audubon Society	New York, USA
National Parks and Wildlife Service, NSW	New South Wales
National Wildlife Federation, USA	Washington DC, USA
Nature Conservation Council of NSW	New South Wales
New South Wales Institute of Technology Students' Association	New South Wales
North Yorke Peninsula Conservation Historical Society	South Australia
Oleofina	Bruxelles, Belgium
Oregonians Co-operating to Protect Whales	Oregon, USA
Orobis Australia Pty Ltd	New South Wales
Paykel Oils and Chemicals Pty Ltd	New South Wales
Pearsall Chemical Corporation	Texas, USA
Port Hacking Divers	New South Wales
Prevention of Cruelty to Animals and Plants	Western Australia
Project Jonah, Sydney	New South Wales
Project Jonah, Melbourne	Victoria

Project Jonah, Brisbane	Queensland
Project Jonah (New Zealand) Inc.	Auckland, New Zealand
Project Jonah South Australia	South Australia
Queensland Conservation Council Inc.	Queensland
Queensland Wildlife Preservation Society	Queensland
Rare Animal Relief Effort Inc.	New York, USA
RSPCA, New South Wales	New South Wales
RSPCA, Victoria	Victoria
RSPCA, Western Australia	Western Australia
Salamanca Company	Tasmania
Sandoz Australia Pty Ltd	Victoria
Sierra Club Office of International Environment Affairs	New York, USA
Sigma Chemicals Pty Ltd	Western Australia
South Australian Government (Minister for the Environment)	South Australia
South Australian Young Liberals	South Australia
South Gippsland Conservation Society	Victoria
Steel Improvement (NSW) Pty Ltd	New South Wales
Stephenson Bros Ltd	West Yorkshire, United Kingdom
Sydney Unitarian Church	New South Wales
Tasmanian Environment Centre	Tasmania
United States Department of Commerce	Washington DC, USA
Victorian Chemical Company Pty Ltd	Victoria
Werner G. Smith, Inc.	Ohio, USA
Western Australian Government	Western Australia
Whale Center	California, USA
Whale Protection Fund	Washington DC, USA

Whalers Haven	South Australia
Wildlife Preservation Society of Australia	New South Wales
Wildlife Preservation Society of Queensland Inc.	Queensland
Wool Textile Manufacturers of Australia	Victoria
World Wildlife Fund	Switzerland
Victorian State Government (Ministry for Conservation)	Victoria

Individuals

Dr K.R. Allen	New South Wales
Mr J.L. Bannister	Western Australia
Mr J. Barzdo	London, United Kingdom
Dr P. Best	Capetown, South Africa
Dr L.K. Boerema	Rome, Italy
Mr M. Bossley	South Australia
Mr M. Brenner	Sarasota, USA
Dr M.M. Bryden	Cambridge, United Kingdom
Mr & Mrs T.J. Butcher	Victoria
Dr R.G. Chittleborough	Western Australia
Mr & Mrs A.E. Cooper	Western Australia
Mrs M.J. Couchman	Victoria
Ms S.A. Earle	San Francisco, USA
Ms J. Esposito	New South Wales
Dr I.A. Furzer	New South Wales
Mr W. Godfrey-Smith	Australian Capital Territory
Mr S.H. Goodall	New South Wales
Mrs J.W. Green	Ohio, USA

Dr E.R. Guiler	Tasmania
Dr J.A. Gulland	Rome, Italy
Dr J. Hatch	South Australia
Dr G.E. Heinsohn	Queensland
Professor L.M. Herman	Hawaii, USA
Dr S. Holt	Rome, Italy
Ms B. Hutton	Victoria
Dr M. Jacobs	New York, USA
Mr & Mrs J. & S.M. Jedryk	New South Wales
Drs D. & C. Jeffs	Western Australia
Professor H.J. Jerison	Los Angeles, USA
Mr R.S.L. Jones	New South Wales
Mr L.T.W. Kennedy	New South Wales
Mr M.G. Kennedy	New South Wales
Mr R.P. Kenny	Queensland
Mr R.H. Knott	New South Wales
Mr P.G. Laird	New South Wales
Ms E.L. Laube	New South Wales
Mr & Mrs C. Leitch	Western Australia
Dr J.C. Lilly	California, USA
Mr R. Loh	New South Wales
Ms J. McIntyre	California, USA
Ms H. Marsh	Queensland
Professor R.M. May	New Jersey, USA
Mr M.C. Mercer	Ontario, Canada
Mr N.E. Milward	Queensland
Dr P. Morgane	Massachusetts, USA
Mr S. Myers	New South Wales

Mr F.L. Nelki	Queensland
Professor M. Nishiwaki	Japan
Dr R. Paterson	Queensland
Mr T.S. Pearce	Victoria
Dr G. Pilleri	Switzerland
Mr D.S. Poignand	Australian Capital Territory
Mr A. Reding	New Jersey, USA
Mr & Mrs I. Robertson	Australian Capital Territory
Ms E. Robinson	Australian Capital Territory
Mr F.D. Robson	Napier, New Zealand
Mr K. Schoeffel	Victoria
Dr E.W. Shallenberger	Hawaii, USA
Mrs J. Sharma	South Australia
Mr R. Shipton, MP	Australian Capital Territory
Professor P. Singer	Victoria
Dr G.L. Small	New York, USA
Mr G. Synnot	Victoria
Mr D.J. Tranter	New South Wales
Mr P. Waters	Australian Capital Territory
Dr W.A. Watkins	Massachusetts, USA
Mr L.H. Watt, MLA	Western Australia
Mrs H.L. Weaver	Surrey, United Kingdom
Mr M. Weinrich	New York, USA
Mr D. Wighton	Edmonton, Canada
Mr H. Wildes	Victoria
Mr E. Worrell	New South Wales
Mrs I. Zahra	New South Wales

APPENDIX 2: EXPERTS CONSULTED DURING OVERSEAS VISIT

Dr R.G. Anderson   Chevron Chemical Corporation
                         California, USA

Mr B. Applebaum   Head
                         Canadian Delegation to preparatory
                             conference on revision of the International
                             Convention for the Regulation of Whaling,
                             Copenhagen, July 1978

Mr I.G. Barr   Highgate and Job Ltd
                         Paisley, Scotland

Dr J.R. Beddington   Department of Biology
                         University of York
                         York, United Kingdom

Dr P.B. Best   Sea Fisheries Branch
                         South African Department of Industries
                         Capetown, South Africa

Dr S. Bunnell   California, USA

Dr D.G. Chapman   Dean
                         College of Fisheries
                         University of Washington
                         Washington, USA

Dr C.W. Clark   Department of Mathematics
                         University of British Columbia
                         Canada

Mr R. Eisenbud	Member USA Delegation to International Whaling Commission
Ms P. Fox	Member USA Delegation to International Whaling Commission
Dr R. Gambell	Secretary International Whaling Commission Cambridge, United Kingdom
Mr E. Greenberg	Member USA Delegation to International Whaling Commission
Professor R.J. Harrison	Professor of Anatomy Cambridge University Cambridge, United Kingdom
Mr W.F.W. Hendry	Highgate and Job Ltd Paisley, Scotland
Dr S. Holt	Adviser on Marine Affairs Fisheries Department, FAO Rome, Italy
Dr M.S. Jacobs	Associate Professor and Acting Chairman Department of Pathology New York University College of Dentistry and Cetacean Neuroanatomist Osborn Laboratories of Marine Sciences New York Aquarium

Professor H.J. Jerison	Department of Psychiatry Centre for the Health Services University of California, USA

Dr R.M. Laws	Director of the British
	    Antarctic Survey and
	    The Sea Mammal Research
	    Unit of the National
	    Environment Research Council
	Cambridge, United Kingdom

Dr J.C. Lilly	The Human/Dolphin Foundation California, USA
Mr C.J. Marsh	Department of Industry London, United Kingdom
Dr W.R. Martin	Member Canadian Delegation to International Whaling Commission
Mr M.C. Mercer	Canadian Commissioner to International Whaling Commission
Dr E. Mitchell	Arctic Biological Station Quebec, Canada
Dr P.J. Morgane	Senior Scientist Worcester Foundation for Experimental Biology Massachusetts, USA
Dr W.R. Payne	New York Zoological Society Lincoln, USA
Ms K. Pryor	New York, USA

Dr V.B. Scheffer   Consultant
                        Marine Mammal Commission
                        Washington, USA

Dr L. Talbot   Senior Scientist
                        Council of Environment Quality
                        Executive Office of the President
                        Washington DC, USA

APPENDIX 3: WITNESSES AT PUBLIC HEARINGS

A. ORGANISATIONS WHOSE REPRESENTATIVES GAVE EVIDENCE AT PUBLIC HEARINGS OF THE INQUIRY

Organisation	Representative
Albany Chamber of Commerce	Mr R.G. Malta
Australian Catholic Study Circle for Animal Welfare	Miss M. Craig
Australian Conservation Foundation	Dr F. Harmon Dr J.G. Mosley Mr A. Tingay
Australian Workers Union of Western Australia	Mr G.A. Barr
Cheynes Beach Holdings Ltd	Mr A.G. Cruickshank Mr J.W. Saleeba Mr V.M. Walters
Commonwealth Government Agencies and Departments	Mr G. Anderson Mr G.C. Evans Mr A.V. Folkard Mr G.C. Gorrie Mr E.A. Purnell-Webb
Friends of the Earth (Carlton)	Miss B.J. Hutton
Friends of the Earth (NSW)	Ms S. Adam Mr P. Power
Friends of the Earth (WA)	Mr P. Brotherton

Greenpeace Australia	Miss J. Adams
Maritime Workers Union	Mr J.C. Wells
Nature Conservation Council of NSW	Mr J.K. Hibberd
Project Jonah	Mr A.E.J. Black
	Mr A.I. Gregory
	Mr P. Kaye
	Mr R.J. Lomax
	Mr R.E. McMillan
	Mr G.W. Smith
	Mr K.D.
	Mr B.W. Tilley
	Mr G. Waugh
Sigma Chemicals Pty Ltd	Mr D. Brown
South Coast Licensed Fishermen's Association of Albany	Mr R.L. Heberle
State Government, New South Wales National Parks and Wildlife Service of New South Wales	Mr W.S. Steel
State Government, Victoria Victorian Fisheries and Wildlife Division - Ministry for Conservation	Mr K.J. Street
State Government, Western Australia Department of Industrial Development	Mr S.G. Grocott
Department of Fisheries and Wildlife	Mr N.W. McLaughlan
The Office of Regional Administration, Great Southern Region	Mr K.R. Marshall

B: INDIVIDUALS WHO GAVE EVIDENCE AT PUBLIC HEARINGS OF THE INQUIRY

Dr K.R. Allen
Mr J.L. Bannister
Mr H.L. Berg
Dr J.C. Clarke
Mr D.D. Clayton
Dr W.H. Dawbin
Mr T.J. Donoghue
Dr I.A. Furzer
Mr M. Greenwood

Dr E.R. Guiler
Dr S.J. Holt
Dr M.S. Jacobs
Dr G.P. Kirkwood
Ms J.A.E.J. Robinson
Mr I.D. Saulwick
Mr P. Waters
Mr J. Watt
Mr H. Wildes

APPENDIX 4 : PRESS RELEASE OF THE CHEYNES BEACH
WHALING CO (1963) PTY LTD, 31 JULY 1978

Whaling operations at Albany are to end - in the near future.

The executive director of Cheynes Beach Holdings Ltd, Mr John Saleeba announced this today at Albany at the opening of the Inquiry into Whales and Whaling.

In a separate statement to shareholders, the chairman, Mr R.L. Hunt, said that there had been a sharp downturn in the demand for crude sperm oil. The directors believed operations this year would result in a substantial loss and it was unlikely there would be any profit in whaling in 1979.

The Stock Exchange was notified of the company's intentions today.

Mr Hunt said that the recent visit to Europe and the United Kingdom by two of the company's directors confirmed that there was a serious move from filtered sperm oil to alternatives. The reason for the move away from sperm oil was the doubt of continuity of supply from Australia, being an unforseen effect of the Public Inquiry of which the first item of reference is whether Australian whaling should continue or cease. Hitherto, Australia had fulfilled a substantial proportion of the Free World demand.

In his statement, Mr Hunt said that from the early part of this year, the company's usual buyers had shown extreme reluctance to commit themselves to forward buying.

So far, one sale of 1,000 tonnes had been made out of this year's operations, but sales were slow and prices were below the cost of production.

Mr Hunt went on: 'Historically, the sperm oil market has always been of a volatile nature but it is difficult to foresee any satisfactory upturn in the near or medium term future.

'The directors believe that whaling operations for 1978 will result in a substantial loss.

'Assuming we receive a whaling licence from the Australian Government for 1979, our quota for that year compared with 1978 has been drastically cut: males by 25% and the less productive females by 10%, resulting in an estimated drop in production of sperm oil of 1,000 tonnes.

'Although certain direct operating costs should fall in line with the reduced production, it is not possible to cut most overhead charges proportionately and thus total costs per tonne for oil must rise very substantially.

'Oil revenue comprises nearly 80% of our total whale product sales and it therefore does not seem feasible to forecast any likelihood of profitable whaling operations in 1979.'

Mr Hunt said in view of the facts and in the best interests of the shareholders the board had decided whaling operations must end in the near future.

The decision was taken with regret and reluctance.

The actual date of cessation would be decided after consideration had been given to forward commitments to the Australian industry and to dealing fairly with the employees of the company.

Mr Hunt said that the closing of the whaling operations did not necessarily mean that the processing plant could not be used for other purposes. He said that a large portion of the plant would need little alteration to cover the processing of fish meal, a high protein product for which there was a world demand.

'Government assistance will be necessary', he said. 'Firstly, an alteration to our lease clauses permitting use of Station land for purposes other than whaling, and secondly, some financial guarantees to enable any viable venture to be established.

'Shareholders may be assured that the Directors will be investigating and pursuing to the utmost any project that will lead to the utilisation of the Albany plant in a profitable enterprise'.

In a separate statement in Albany, Mr Saleeba said that the closing of the whaling operations after more than a quarter of a century was based solely on economic considerations.

'The shutting down doesn't mean that the whaling will not continue, the Australian quota will merely go back into the international pool and other whaling nations will take our share.'

APPENDIX 5: CLASSIFICATION OF THE ORDER CETACEA

The taxonomy of the order Cetacea varies considerably among whale biologists (cetologists). There may be differences of opinion on certain of the groupings, and scientific nomenclature, of several of the cetacean species. The table below is based largely on the taxonomy found in a technical report A list of the marine mammals of the world (3rd edition) prepared by D.W. Rice in 1977 and published by the National Oceanic and Atmospheric Administration of the United States Department of Commerce. In addition the works of Bunnell (1974) and Walker (1975) were used. Divisions within species (that is sub-species, or in some cases races) are recognised for some species but have not been included here.

Cetacea which have been found in the waters around Australia are marked with an asterisk(*).

Sub-order Mysticeti (baleen whales - baleen plates or 'sieves' in mouth, no teeth, double blowhole, symmetrical skull)
    Family Balaenopteridae (rorqual whales - refers to series of grooves, or folds, along throat; have distinct dorsal fin)
        Genus Balaenoptera
- * Balaenoptera acutorostrata - minke whale (up to 9 metres)
- * Balaenoptera borealis - sei whale (18 metres)
- * Balaenoptera edeni - Bryde's whale (15 metres)
- * Balaenoptera musculus - blue whale (30 metres)
- * Balaenoptera physalus - fin whale (25 metres)

        Genus Megaptera
- * Megaptera novaeangliae - humpback whale (14 metres)

    Family Balaenidae (right whales - long baleen plates, no dorsal fin or throat grooves)
        Genus Balaena
            Balaena mysticetus - bowhead or Greenland right whale (18 metres)

Genus Eubalaena[1]
* Eubalaena glacialis - right whale or black right whale (16 metres)
Genus Caperea
* Caperea marginata - pygmy right whale (6 metres)
Family Eschrichtiidae
Genus Eschrichtius
Eschrichtius robustus - gray whale (15 metres)

Sub-order Odontoceti (toothed whales, simple conical teeth, single blowhole, asymmetrical skull)

Family Physeteridae (sperm whales - blowhole on left side of head, prominent teeth on lower jaw)
Genus Physeter
* Physeter catodon - sperm whale (18 metres)
Family Kogiidae (pygmy sperm whales)
Genus Kogia
* Kogia breviceps - pygmy sperm whale (4 metres)
* Kogia simus - dwarf sperm whale (2.5 metres)
Family Ziphiidae (beaked whales)
Genus Berardius
Berardius bairdii - Baird's beaked whale (13 metres)
* Berardius arnuxii - Arnoux's beaked whale (11 metres)
Genus Hyperoodon
Hyperoodon ampullatus - northern bottlenosed whale (9 metres)
* Hyperoodon planifrons - southern bottlenosed whale (9 metres)
Genus Ziphius
* Ziphius cavirostris - Cuviers' beaked whale (6 metres)
Genus Tasmacetus
Tasmacetus shepherdi - Shepherd's beaked whale (6 metres)
Genus Mesoplodon
* Mesoplodon mirus - True's beaked whale (5 metres)
* Mesoplodon hectori - Hector's beaked whale (5 metres)
* Mesoplodon layardii - strap-toothed whale (6 metres)

---

(1) Some whale taxonomists consider that the genus Eubalaena contains two distinct species, as follows:
Eubalaena glacialis - northern right whale
* Eubalaena australis - southern right whale

 * Mesoplodon grayi - camperdown whale (4 metres)
  Mesoplodon bidens - Sowerby's beaked whale (5 metres)
  Mesoplodon ginkgodens - Japanese beaked whale (6 metres)
  Mesoplodon carlhubbsi - Hubb's beaked whale (5 metres)
  Mesoplodon stejnegeri - Stejneger's beaked whale (6 metres)
 * Mesoplodon bowdoini - Andrew's beaked whale (4.5 metres)
 * Mesoplodon densirostris - Blainville's beaked whale (4.5 metres)
  Mesoplodon europaeus - Antillean beaked whale (4 metres)
 Genus Indopacetus
* Indopacetus pacificus - Pacific beaked whale (6 metres)
Family Monodontidae (white whales)
 Genus Delphinapterus
  Delphinapterus leucas - beluga (5.5 metres)
 Genus Monodon
  Monodon monoceros - narwhal (4.5 metres)
Family Platanistidae (river or freshwater dolphins)
 Genus Platanista
  Platanista gangetica - susu or Ganges River dolphin (2.5 - 3 metres)
  Platanista indi - Indus River dolphin (2.5 metres)
 Genus Inia
  Inia geoffrensis - Amazon River dolphin (3 metres)
 Genus Lipotes
  Lipotes vexillifer - Chinese lake dolphin (2.4 metres)
 Genus Pontoporia
  Pontoporia blainvillei - La Plata dolphin (1.5 metres)
Family Delphinidae (dolphins - usually have a beak, characteristic triangular dorsal fin)
 Genus Lissodelphis
  Lissodelphis borealis - northern right whale dolphin (2.5 metres)
 * Lissodelphis peronii - southern right whale dolphin (2.3 metres)
 Genus Delphinus
 * Delphinus delphis - common dolphin (2.3 metres)
 Genus Lagenorhynchus
  Lagenorhynchus obliquidens - Pacific white-sided dolphin (2 - 2.5 metres)

    Lagenorhynchus albirostris - white-beaked dolphin (2.5 - 3 metres)
* Lagenorhynchus obscurus - dusky dolphin (2 metres)
    Lagenorhynchus acutus - Atlantic white-sided dolphin (2.7 metres)
    Lagenorhynchus thicolea - Falkland Island dolphin (2.5 metres)
* Lagenorhynchus cruciger - hour-glass dolphin (1.8 metres)
    Lagenorhynchus australis - blackchin dolphin (2.2 metres)

Genus Tursiops

* Tursiops truncatus - Atlantic bottlenose dolphin (3 - 3.5 metres)
    Tursiops gilli - Pacific bottlenose dolphin (3.5 metres)

Genus Grampus

* Grampus griseus - Risso's dolphin (4.3 metres)

Genus Lagenodelphis

* Lagenodelphis hosei - Sarawak dolphin (2.5 metres)

Genus Feresa

* Feresa attenuata - pygmy killer whale (2.5 metres)

Genus Cephalorhynchus

    Cephalorhynchus commersonii - Commerson's dolphin (1.6 metres)
    Cephalorhynchus hectori - Hector's dolphin (1.8 metres)
    Cephalorhynchus heavisidii - Heaviside's dolphin (1.5 metres)
    Cephalorhynchus eutropia - white-bellied dolphin (1.5 metres)

Genus Orcinus

* Orcinus orca - killer whale (9 metres)

Genus Pseudorca

*Pseudorca crassidens - false killer whale (5.5 metres)

Genus Orcaella

* Orcaella brevirostris - Irrawaddy dolphin (2.1 metres)

Genus Globicephala

* Globicephala melaena - longfin pilot whale (6 metres)
* Globicephala macrorhynchus - shortfin pilot whale (6 metres)

Genus Peponocephala

* Peponocephala electra - broad-beaked dolphin or many-toothed blackfish (2.8 metres)

Family Stenidae (other dolphins)

Genus Steno

* Steno bredanensis - rough-toothed dolphin (2.5 metres)

Genus Sotalia
    Sotalia fluviatilis - Bouto dolphin (1.2 metres)
    Sotalia brasiliensis - Rio de Janeiro dolphin (1.5 metres)
    Sotalia guianensis - Guiana River dolphin (1.5 metres)
Genus Sousa
    Sousa chinensis - Chinese white dolphin (1.5 metres)
* Sousa borneensis - Borneo white dolphin (2 metres)
* Sousa lentiginosa - freckled dolphin (2.5 metres)
* Sousa plumbea - plumbeous dolphin (2.4 metres)
    Sousa teuszii - Cameroon dolphin (2.3 metres)
Genus Stenella
* Stenella coeruleoalba - blue or striped dolphin (2.7 metres)
* Stenella longirostris - spinning dolphin (2.1 metres)
* Stenella attenuata$^{(1)}$ - bridled dolphin (2.1 metres)
* Stenella plagiodon - spotted dolphin (2.2 metres)
Family Phocoenidae (porpoises - lacking in prominent beak, dorsal fin low and triangular)
    Genus Phocoena
        Phocoena phocoena - harbor porpoise (1.8 metres)
        Phocoena dioptrica - spectacled porpoise (1.8 metres)
        Phocoena sinus - Gulf of California porpoise (1.5 metres)
        Phocoena spinipinnis - black porpoise (1.5 metres)
    Genus Phocoenoides
        Phocoenoides dalli - Dall's porpoise (2.2 metres)
    Genus Neophocaena
        Neophocaena phocaenoides - black finless porpoise (1.5 metres)

---

(1) Include S. graffmani, S. dubia and S. frontalis in this complex.

APPENDIX 6: INTERNATIONAL WHALING COMMISSION
DIVISIONS AND AREAS IN THE SOUTHERN HEMISPHERE

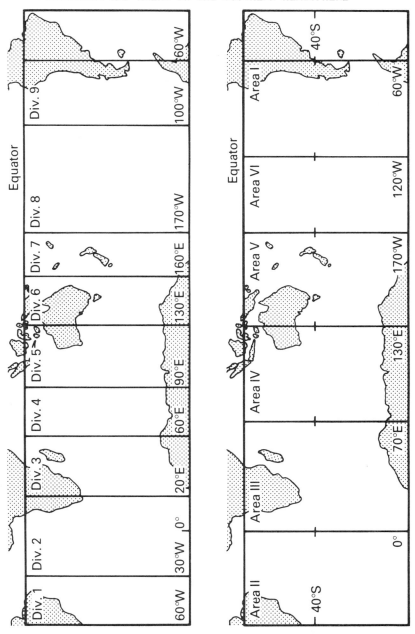

APPENDIX 7: AUSTRALIAN WHALING LEGISLATION AND LICENCES

- (i) Whaling Act 1960
- (ii) Statutory Rules 1975 No. 105
- (iii) Licence to use ship or aircraft as whale catcher
- (iv) Licence to use a ship as a factory ship or a factory as a land station

WHALING ACT 1960

Reproduced from Acts of the Parliament 1901 - 1973.

The Act is shown as in force at 30 November 1978.

Note: Before its repeal, section 4 provided as follows:
'The Whaling Act 1935 and the Whaling Act 1948 are repealed'.

COMMONWEALTH OF AUSTRALIA

WHALING ACT 1960-1973
TABLE OF PROVISIONS
PART I—PRELIMINARY

Section
1. Short title
2. Commencement
3, 4. (Repealed)
5. Interpretation
6. Act to bind Crown
7. Extra-territorial operation of Act
8. Application of Act to State territorial waters
9. Delegation

PART II—REGULATION OF WHALING

10. Prohibition of certain acts by notice
11. Licences
12. Conditions of licences
13. Whaling inspectors
14. Powers of officers

PART III—OFFENCES

15. Contravention of notices
16. Use of unlicensed ships and factories
17. Breach of conditions of licence
18. Possession, &c., of whales illegally killed
19. Forfeiture of equipment, &c.
20. Unlicensed ships entering Australia
21. Remuneration of gunners and crew, &c.
22. Obstruction of officers, &c.
23. Punishment of offences

PART IV—RESEARCH AND DEVELOPMENT

24. Exploratory operations
25. Investigations
26. Permit for scientific purposes

PART V—MISCELLANEOUS

27. Jurisdiction of courts
28. Regulations

WHALING ACT 1960-1973

An Act relating to Whaling.

PART I—PRELIMINARY

1. This Act may be cited as the *Whaling Act* 1960-1973.[1]

Short title.
Short title amended: No. 32, 1918, s. 2.

2. This Act shall come into operation on a date to be fixed by Proclamation.[1]

Commencement.

*   *   *   *   *   *   *   *

Sections 3 and 4 repealed by No. 216, 1973, s. 3.

5. (1) In this Act, unless the contrary intention appears—

"Australia" includes the Territories;

"Australian waters" means—
   (a) Australian waters beyond territorial limits;
   (b) the waters adjacent to a Territory and within territorial limits; and
   (c) the waters adjacent to a Territory, being a Territory that is not part of the Commonwealth, and beyond territorial limits;

"baleen whale" means a whale other than a toothed whale;

"blue whale" means a whale of the genus and species *Balaenoptera musculus* or *Sibbaldus musculus*, that is to say, a whale of the kind known by the name of blue whale, Sibbald's rorqual or sulphur bottom;

"factory" means a factory situated in Australia;

"factory ship" means a ship in or on which whales are treated, whether wholly or in part, but does not include a ship used solely for freezing or salting the meat and entrails of whales intended for human consumption or use as animal food;

"fin whale" means a whale of the genus and species *Balaenoptera physalus*, that is to say, a whale of the kind known by the name of common finback, common rorqual, finback, finner, fin whale, herring whale, razorback or true fin whale;

"gray whale" means a whale of the genus and species *Rhachianectes glaucus*, that is to say, a whale of the kind known by the name of California gray, devil fish, gray back, gray whale, hard head, mussel digger or rip sack;

Interpretation.
Sub-section (1) amended by No. 216, 1973, s. 3.

"humpback whale" means a whale of the genus and species *Megaptera nodosa* or *Megaptera novaeangliae*, that is to say, a whale of the kind known by the name of bunch, hump whale, humpback, humpback whale, humpbacked whale or hunchbacked whale;

"land station" means a factory at which whales are treated;

"licence" means a licence granted under this Act;

"master", in relation to a ship, includes any person in charge of the ship;

"minke whale" means a whale of the genus and species *Balaenoptera acutorostrata, Balaenoptera davidsoni* or *Balaenoptera huttoni*, that is to say, a whale of the kind known by the name of lesser rorqual, little piked whale, minke whale, pike-headed whale or sharp headed finner;

"officer" means—

(a) a person permanently or temporarily employed in the Public Service of the Commonwealth or of a Territory or by an authority of the Commonwealth and authorized by the Secretary to perform the duties of an officer under this Act;

(b) a person permanently or temporarily employed in the Public Service of a State and authorized by the Secretary to perform the duties of an officer under this Act in pursuance of an arrangement between the Commonwealth and the State;

(c) a member of the police force of the Commonwealth or a State or Territory; or

(d) a member of the Defence Force;

"pilot", in relation to an aircraft, means the person in charge or command of the aircraft;

"port" includes any place in or at which ships can obtain shelter or ship and unship goods;

"right whale" means a whale of the genus and species *Balaena mysticetus, Eubalaena glacialis, Eubalaena australis* or *Neboalaena marginata*, that is to say, a whale of the kind known by the name of Arctic right whale, Atlantic right whale, Biscayan right whale, bowhead, great polar whale, Greenland right whale, Greenland whale, Nordkaper, North Atlantic right whale, North Cape whale, Pacific right whale, pigmy right whale, Southern pigmy right whale or Southern right whale;

"sei whale" means a whale of the genus and species *Balaenoptera borealis*, that is to say, a whale of the kind known by the name of coalfish whale, pollack whale, Rudolphi's rorqual or sei whale, and includes a whale of the genus and species *Balaenoptera brydei*, that is to say, a whale of the kind known by the name of Bryde's whale;

"ship" includes every kind of vessel;

"sperm whale" means a whale of the genus and species *Physeter catadon*, that is to say, a whale of the kind known by the name of cachalot, pot whale, sperm whale or spermacet whale;

"take" in relation to whales, means take, catch or capture, and "taking", in relation to whales, has a corresponding meaning;

"the Convention of 1946" means the International Convention for the Regulation of Whaling signed at Washington on the second day of December, One thousand nine hundred and forty-six;

"the International Whaling Conventions" means the Convention of 1946 and the Protocol to that Convention dated the nineteenth day of November, One thousand nine hundred and fifty-six, and includes any amendment of the Schedule to the Convention of 1946 made, whether before or after the commencement of this Act, in pursuance of Article V of that Convention (including that Article as amended by the Protocol), being an amendment that has become effective with respect to the Government of the Commonwealth;

"the owner", in relation to a ship or aircraft, includes—

(a) every person who is a co-owner of the ship or aircraft or of any part of or share in the ship or aircraft; and

(b) where a company or body corporate owns the ship or aircraft, or is a co-owner of the ship or aircraft or of a part of or share in the ship or aircraft—a person who is the manager or secretary of that company or body corporate;

"the Secretary" means the Secretary to the Department of Primary Industry;

"toothed whale" means a whale that has teeth in the jaws;

"treating", in relation to whales, includes any operation of cutting up, or of extracting oil, whalebone or other products from, the carcases of whales, and "treat" and "treated", in relation to whales, have corresponding meanings;

"waters to which this Act applies" means—

(a) Australian waters; and

(b) subject to section eight of this Act, all other waters;

"whale" means—

(a) a blue whale, fin whale, gray whale, humpback whale, right whale, sei whale or other baleen whale;

(b) a sperm whale; or

(c) any other whale of a prescribed kind;

"whale catcher" means a ship (other than a factory ship) used for the purpose of hunting, taking, killing, towing, holding on to or scouting for whales or an aircraft used for such a purpose.

(2) For the purposes of this Act, a ship or aircraft shall be deemed to be under the jurisdiction of the Commonwealth if—

Whaling Act 1960-1973

(a) it is registered in Australia;
(b) its operations are based on a port or place in Australia; or
(c) it is within the territorial limits of the Commonwealth or of a Territory and is not a public ship or aircraft of a country other than Australia that is neither employed for the purposes of whaling nor otherwise employed in commercial operations.

Act to bind Crown.

6. This Act binds the Crown in right of the Commonwealth or of a State and any authority constituted by or under a law of the Commonwealth, a State or a Territory.

Extra-territorial operation of Act.

7. This Act applies both within and without the Commonwealth and extends to all the Territories.

Application of Act to State territorial waters.

8. (1) A reference in this Act to waters to which this Act applies shall be read as not including a reference to waters that are territorial waters of a State unless a Proclamation under the next succeeding sub-section is in force in respect of those waters.

(2) The Governor-General may, by Proclamation, declare that this Act applies in respect of the Territorial waters of a State or a specified part of those territorial waters.

Delegation.

9. (1) The Minister or the Secretary may, either generally or in relation to a matter or class of matters and either in relation to the whole or a part of Australia or in relation to the waters to which this Act applies or a part of those waters, by writing under his hand, delegate all or any of his powers and functions under this Act or the regulations, except this power of delegation.

(2) A power or function so delegated may be exercised or performed by the delegate in accordance with the instrument of delegation.

(3) A delegation under this section is revocable at will and does not prevent the exercise of a power or the performance of a function by the Minister or the Secretary.

PART II—REGULATION OF WHALING

Prohibition of certain acts by notice.

10. (1) Subject to sub-section (5) of this section, the Minister may, by notice published in the *Gazette*, prohibit, either at all times or during a period specified in the notice—
(a) the taking or killing of whales, or whales of a species, kind or sex specified in the notice;
(b) the taking or killing of whales, or whales of a species, kind or sex specified in the notice, not exceeding a size so specified; or
(c) the taking or killing of whales, or whales of a species, kind or sex specified in the notice, by a method or equipment so specified.

(2) A notice under the last preceding sub-section applies to the taking or killing of whales in any waters to which this Act applies unless the notice is expressed to apply only in relation to a part of those waters.

(3) The power conferred by virtue of paragraph (a) of sub-section (1) of this section extends to prohibiting the taking or killing of female whales, or female whales of a particular species or kind, when accompanied by calves or suckling whales.

(4) A notice under this section may provide for exceptions to, and exemptions from, the prohibition contained in the notice and such an exception or exemption has effect subject to such conditions, if any, as are specified in the notice.

(5) The powers conferred on the Minister by this section are, in relation to the taking or killing of whales in waters other than Australian waters, exercisable only to the extent necessary to give effect to the International Whaling Conventions.

11. (1) Subject to this section, the Secretary may, in his discretion, grant to a person, being the owner or charterer of a ship or aircraft, a licence to use that ship or aircraft as a whale catcher in, or in and over, the waters to which this Act applies or such of those waters as are specified in the licence. *Licences.*

(2) Subject to this section, the Secretary may, in his discretion, grant to a person, being the owner or charterer of a ship or the occupier of a factory, a licence to use that ship as a factory ship, or to use that factory as a land station, as the case may be, for the treating of whales taken or killed in the waters to which this Act applies or such of those waters as are specified in the licence.

(3) A licence shall not be granted under either of the last two preceding sub-sections in respect of a ship that is not registered in Australia unless the ship is duly authorized by the Government of the country whose flag she flies to engage in taking and killing whales or in treating whales, as the case requires.

(4) Subject to this section, a licence granted under this section remains in force for such period, not exceeding five years, as is specified in the licence.

(5) The Secretary may, in his discretion, on the application of the holder of a licence granted under this section and of another person as proposed transferee, transfer the licence to that other person.

(6) Such fees, if any, as are prescribed are payable in respect of the grant of a licence, or the transfer of a licence, under this section.

(7) Where—
(a) the holder of a licence has been convicted of an offence against this Act or the regulations; or

Whaling Act 1960-1973

(b) the Secretary is satisfied that there has been a contravention of, or failure to comply with, a condition of a licence granted under this section,

the Secretary may cancel the licence.

(8) The Secretary may require an applicant for a licence or the holder of a licence to give security to his satisfaction for compliance with the conditions of the licence and with the requirements of this Act and the regulations and, if the applicant or holder fails to give that security, may refuse to grant the licence or may cancel the licence, as the case requires.

(9) A register showing the licences granted under this section and in force from time to time shall be kept at such place as the Minister directs.

Conditions of licences.
12. (1) A licence granted under the last preceding section is subject to such conditions as are specified in the licence.

(2) The conditions subject to which a licence is granted shall include such conditions as the Secretary considers necessary to give effect to the International Whaling Conventions, including conditions to ensure that there will be maximum utilization of the carcases of whales taken or treated by the holder of the licence.

(3) Conditions that relate to the taking or killing of whales, or to whales taken or killed, in waters other than Australian waters shall not be specified in a licence except for the purpose of giving effect to the International Whaling Conventions.

Whaling inspectors.
13. (1) The Secretary may appoint persons to be whaling inspectors for the purposes of this section.

(2) Subject to the next succeeding sub-section, not less than two whaling inspectors shall be maintained on board each factory ship in respect of which a licence under this Act is in force.

(3) If a factory ship not registered in Australia carries whaling inspectors in accordance with the law of the country whose flag she flies, the Secretary may, by instrument in writing, exempt the ship from the operation of the last preceding sub-section.

(4) A whaling inspector, or, if the Secretary thinks necessary, two or more whaling inspectors, shall be maintained at each land station in respect of which a licence under this Act is in force.

(5) A whaling inspector maintained on board a factory ship or at a land station is entitled to remain on board the ship or upon the station premises and to be present at all operations in connexion with the treating of whales on board the ship or at the station.

Whaling Act 1960-1973

(6) The master of a ship on board which a whaling inspector is maintained under this section, and the occupier of a land station at which a whaling inspector is so maintained, shall provide the inspector with reasonable accommodation and subsistence.

Penalty: Five hundred dollars.

Amended by No. 93, 1966, s. 3.

(7) The Commonwealth shall pay to the owner, charterer or master of such a ship, or the occupier of such a land station, in respect of each whaling inspector who is provided with accommodation and subsistence on board the ship or at the station in pursuance of this section, such amount for each day on which that accommodation and subsistence is provided as is fixed by or under the regulations.

14. (1) For the purposes of this Act, an officer may—

Powers of officers.

(a) board a ship or aircraft under the jurisdiction of the Commonwealth which, or which he has reason to believe—
  (i) is a factory ship or a whale catcher; or
  (ii) has been, is being or is intended to be used for a purpose for which a factory ship or a whale catcher is used;
(b) enter a land station, or any premises which he has reason to believe have been used, are being used or are intended to be used for treating whales;
(c) inspect a ship, aircraft, land station or premises which he has boarded or entered in pursuance of this section and the plant and equipment in or on the ship, aircraft, land station or premises, and examine any whale, part of a whale or whale product in or on the ship, aircraft, land station or premises;
(d) require the master or pilot or a member of the crew (including a gunner) of any such ship or aircraft, or the occupier or any person employed in or in connexion with any such land station or premises, to produce to the officer such licences, records and other documents as the officer considers it necessary to inspect for the purposes of this Act, and, subject to the next succeeding sub-section, to give to the officer such information concerning the ship, aircraft, land station or premises, or the persons on board the ship or aircraft or at the land station or on the premises, as the officer considers necessary for the purposes of this Act;
(e) take copies of, or extracts from, documents produced to him;
(f) seize, take, detain, remove and secure—
  (i) any whale, part of a whale or product of a whale which the officer has reason to believe has been taken or killed in contravention of this Act; and
  (ii) any equipment which the officer has reason to believe has been used in taking or killing a whale in contravention of this Act;

Whaling Act 1960-1973

(g) where the officer has reason to believe that any whale, part of a whale, product of a whale or equipment that he is authorized to seize by virtue of the last preceding paragraph is on board a ship or aircraft under the jurisdiction of the Commonwealth, require the master or pilot of the ship or aircraft to bring the ship or aircraft to a port or place in Australia specified by the officer;

(h) sell any whale, part of a whale or product of a whale seized under this Act;

(i) without warrant, arrest a person who the officer has reason to believe has committed an offence against this Act; and

(j) require a person whom he reasonably suspects of having committed an offence against this Act or the regulations to state his name and place of abode.

(2) A person is not obliged to comply with a requirement under paragraph (d) of the last preceding sub-section in so far as it requires him to give information that might incriminate him.

PART III—OFFENCES

Contravention of notices.
Amended by No. 93, 1966, s. 3.

15. A person shall not do an act prohibited by a notice for the time being in force under section ten of this Act.

Penalty: Two thousand dollars.

Use of unlicensed ships and factories.

16. (1) Subject to the next succeeding sub-section, a ship or aircraft under the jurisdiction of the Commonwealth shall not be used for the purpose of hunting, taking, killing, towing, holding on to or scouting for whales in or over any waters to which this Act applies unless the owner or charterer of the ship or aircraft is the holder of a licence granted under sub-section (1) of section eleven of this Act authorizing the use of that ship or aircraft as a whale catcher in, or in and over, those waters.

(2) The use of a factory ship in respect of which a licence under sub-section (2) of section eleven of this Act is in force for the purpose of holding on to a whale shall be deemed not to be a contravention of the last preceding sub-section.

(3) A ship under the jurisdiction of the Commonwealth or a factory shall not be used for treating whales taken or killed in any waters to which this Act applies unless the owner or charterer of the ship or the occupier of the factory is the holder of a licence granted under sub-section (2) of section eleven of this Act authorizing the use of that ship as a factory ship or the use of that factory as a land station, as the case may be, for the treating of whales taken or killed in those waters.

(4) Where a ship, aircraft or factory is used in contravention of a provision of this section— Amended by No. 93, 1966, s. 3.

(a) in the case of a ship—the owner and the master of the ship, or, if the ship is under charter, the charterer and the master of the ship;

(b) in the case of an aircraft—the owner and the pilot of the aircraft, or, if the aircraft is under charter, the charterer and the pilot of the aircraft; or

(c) in the case of a factory—the manager and the occupier of the factory,

are each guilty of an offence against this Act punishable, upon conviction, by a penalty not exceeding Two thousand dollars and, in addition, by a penalty not exceeding One thousand dollars in respect of each whale proved to have been taken or killed by means of the ship or aircraft, or treated on the ship or at the factory, as the case may be, while the ship, aircraft or factory was used in contravention of a provision of this section.

17. A person who is the holder of a licence under section eleven of this Act shall not contravene, or fail to comply with, a condition of the licence. Breach of conditions of licence. Amended by No. 93, 1966, s. 3.

Penalty: Five hundred dollars.

18. (1) Where a whale, a part of a whale or a product of a whale, being a whale taken or killed in contravention of this Act, is found on a ship under the jurisdiction of the Commonwealth or in a factory or other premises, the owner and the master of the ship, or, if the ship is under charter, the charterer and the master of the ship, or the manager and the occupier of the factory or other premises, as the case may be, are each guilty of an offence against this Act punishable, upon conviction, by a penalty not exceeding Two thousand dollars. Possession, &c., of whales illegally killed. Sub-section (1) amended by No. 93, 1966, s. 3.

(2) It is a defence to a prosecution for an offence under the last preceding sub-section in respect of a whale, a part of a whale or a product of a whale found in a factory or other premises if the defendant proves that he was not aware, and had no reasonable grounds for believing, that the whale was taken or killed in contravention of this Act.

19. Where a person is convicted of an offence against this Act in respect of the taking, killing or treating of a whale in contravention of this Act, the court by which he is convicted may order the forfeiture to the Commonwealth of— Forfeiture of equipment, &c.

(a) any equipment used in contravention of this Act in the taking or killing of the whale; or

(b) the whale, or any part or product of the whale, or the proceeds of the sale of the whale or of any part or product of the whale.

Whaling Act 1960-1973

Unlicensed ships entering Australia.

20. (1) A ship designed and equipped for hunting, taking, killing, towing, holding on to or scouting for whales, or for treating whales, shall not be brought into a port in Australia unless—

(a) the owner or charterer of the ship is the holder of a licence in force under this Act authorizing the use of the ship as a whale catcher or as a factory ship, as the case may be; or

(b) the ship is duly authorized by the Government of the country whose flag she flies to engage in taking whales or in treating whales, as the case may be.

Amended by No. 93, 1966, s. 3.

(2) Where a ship is brought into a port in Australia in contravention of the last preceding sub-section, the owner and master, or, if the ship is under charter, the charterer and master, of the ship are each guilty of an offence against this Act punishable, upon conviction, by a penalty not exceeding Two thousand dollars.

Remuneration of gunners and crew, &c.

21. (1) A person shall not engage another person for employment as—

(a) the master or a gunner or member of the crew of a whale catcher, being a ship, or of a factory ship; or

(b) the pilot or a member of the crew of a whale catcher, being an aircraft,

unless the terms of employment of the person so engaged are such that his remuneration is made dependent to a considerable extent upon such factors as the species, size and yield of whales taken and not merely upon the number of whales taken.

(2) A person shall not engage another person for employment as—

(a) the master or a gunner or member of the crew of a whale catcher, being a ship; or

(b) the pilot or a member of the crew of a whale catcher, being an aircraft,

unless it is a term of employment of the person so engaged that no bonus or other remuneration is payable to him in respect of the taking of a whale that is milk-filled or lactating or the taking of which is prohibited by or under this Act.

(3) The last two preceding sub-sections apply to—

(a) an engagement in Australia; and

(b) an engagement outside Australia for employment on a ship or aircraft that is registered in Australia or the operations of which are based on a port or place in Australia.

Amended by No. 93, 1966, s. 3.

(4) A person shall not pay to—

(a) the master or a gunner or member of the crew of a whale catcher, being a ship, under the jurisdiction of the Commonwealth; or

(b) the pilot or a member of the crew of a whale catcher, being an aircraft, under the jurisdiction of the Commonwealth,

a bonus or other remuneration in respect of the taking of a whale that is milk-filled or lactating or the taking of which is prohibited by or under this Act.

Penalty: Five hundred dollars.

22. A person shall not—

(a) fail to facilitate by all reasonable means the boarding of a ship or aircraft or the entry of a land station or other premises by an officer in pursuance of the powers conferred on him by this Act;

(b) refuse to allow an inspection or examination to be made which is authorized by this Act;

(c) subject to sub-section (2) of section fourteen of this Act, refuse or neglect to comply with a requirement made by an officer under sub-section (1) of that section;

(d) when lawfully required to state his name and place of abode to an officer, state a false name or place of abode to the officer;

(e) when lawfully required by an officer to give information, give false or misleading information to the officer;

(f) assault, resist or obstruct an officer or a whaling inspector in the exercise of his powers under this Act;

(g) impersonate an officer; or

(h) in an application under this Act, make a statement or furnish information which is false or misleading in any particular.

Penalty: One thousand dollars.

Obstruction of officers, &c.
Amended by No. 93, 1966, s. 3.

23. (1) An offence against this Act may be prosecuted either summarily or upon indictment, but an offender is not liable to be punished more than once in respect of the same offence.

Punishment of offences.

(2) In summary proceedings against a person for an offence against this Act, the court shall not impose on that person, in respect of the offence, a penalty exceeding, or penalties exceeding in the aggregate, One thousand dollars.

Amended by No. 93, 1966, s. 3.

PART IV—RESEARCH AND DEVELOPMENT

24. The Secretary may, subject to the directions of the Minister, cause operations to be carried out—

(a) for ascertaining whether whaling in particular Australian waters can be engaged in on a commercial basis; and

(b) for the development of whaling in Australian waters.

Exploratory operations.

Whaling Act 1960-1973

Investigations.

25. The Secretary may, subject to the directions of the Minister, cause investigations to be made into economic matters relating to whaling.

Permit for scientific purposes.

26. (1) The Minister may grant a permit to a person authorizing the taking or killing, or the treating, for purposes of scientific research, subject to such restrictions as to number and such other conditions as are specified in the permit, of whales the taking or killing, or the treating, of which is otherwise prohibited by or under this Act.

(2) A person is not guilty of an offence against this Act or the regulations by reason of anything done by him which he is authorized to do by a permit in force under this section.

PART V—MISCELLANEOUS

Jurisdiction of courts.

27. (1) Subject to the succeeding provisions of this section—

(a) the several courts of the States are invested with federal jurisdiction; and

(b) jurisdiction is conferred on the several courts of the Territories,

with respect to offences against this Act or the regulations.

(2) The jurisdiction invested in or conferred on courts by the last preceding sub-section is invested or conferred within the limits (other than limits having effect by reference to the places at which offences are committed) of their several jurisdictions, whether those limits are as to subject-matter or otherwise, but subject to the conditions and restrictions specified in paragraphs (a), (b) and (c) of sub-section (2) of section thirty-nine of the *Judiciary Act* 1903-1959.

(3) The jurisdiction invested in, or conferred on, a court of summary jurisdiction by this section shall not be judicially exercised except by a Chief, Police, Stipendiary, Resident or Special Magistrate, or a District Officer or Assistant District Officer of a Territory.

(4) The trial on indictment of an offence against this Act, not being an offence committed within a State, may be held in any State or Territory.

(5) Subject to this Act, the laws of a State or Territory with respect to the arrest and custody of offenders or persons charged with offences and the procedure for—

(a) their summary conviction;

(b) their examination and commitment for trial on indictment;

(c) their trial and conviction on indictment; and

(d) the hearing and determination of appeals arising out of any such trial or conviction or out of any proceedings connected therewith, and for holding accused persons to bail apply, so far as they are applicable, to a person who is charged in that State or Territory with an offence against this Act or the regulations.

(6) Except as provided by this section, the *Judiciary Act* 1903-1959 applies in relation to offences against this Act or the regulations.

28. (1) Subject to the next succeeding sub-section, the Governor-General may make regulations, not inconsistent with this Act, prescribing all matters which by this Act are required or permitted to be prescribed, or which are necessary or convenient to be prescribed for carrying out or giving effect to this Act, and, in particular, making provision for or in relation to— Regulations. Sub-section (1) amended by No. 93, 1966.

  (a) the furnishing of statistics in relation to—
    (i) the taking, killing and treating of whales;
    (ii) the production of whale products by factory ships or land stations;
    (iii) the number and classes of persons employed in the taking, killing and treating of whales; and
    (iv) the plant and equipment used in the taking, killing and treating of whales;
  (b) the marking of whales taken by whale catchers and the reporting by whale catchers of particulars concerning whales so taken;
  (c) the keeping of records or information relating to the taking, killing and treating of whales and of particulars of whales taken, killed or treated;
  (d) the manner in which whales are to be measured for the purposes of this Act; and
  (e) the imposing of penalties not exceeding Two hundred dollars for offences against the regulations.

(2) The power to make regulations conferred by the last preceding sub-section is, in relation to the taking or killing of whales, and to whales taken or killed, in waters other than Australian waters, exercisable only to the extent necessary to give effect to the International Whaling Conventions.

NOTE

1. The *Whaling Act* 1960-1973 comprises the *Whaling Act* 1960 as amended by the other Acts specified in the following table:

Whaling Act 1960-1973

Act	Number and year	Date of Assent	Date of commencement
Whaling Act 1960	No. 10, 1960	13 May 1960	11 May 1961 (*see Gazette* 1961. p. 1757)
Statute Law Revision (Decimal Currency) Act 1966	No. 93, 1966	29 Oct 1966	1 Dec 1966
Statute Law Revision Act 1973	No. 216, 1973	19 Dec 1973	31 Dec 1973

Statutory Rules

1975 No. 105

REGULATION UNDER THE WHALING ACT 1960-1973.*

I, THE GOVERNOR-GENERAL of Australia, acting with the advice of the Executive Council, hereby make the following Regulation under the *Whaling Act* 1960-1973.

Dated this fifth day of June, 1975.

JOHN R. KERR
Governor-General.

By His Excellency's Command,

K. S. WRIEDT
Minister of State for Agriculture.

AMENDMENT OF THE WHALING REGULATIONS†

After regulation 3 of the Whaling Regulations the following regulation is inserted:—

"3A. For the purposes of paragraph (c) of the definition of 'whale' in sub-section 5 (1) of the Act, a whale included in a taxon specified in the following table is a whale of a prescribed kind:—

[Sidenote: Whales of a prescribed kind for the purposes of the definition of 'whale' in sub-section 5 (1) of the Act.]

Order	Family	Genus
Odontoceti	Delphinidae	*Cephalorhynchus* *Delphinus* *Globicephala* *Grampus* *Lagenorhynchus* *Lissodelphis* *Orcaella* *Orcinus* *Peponocephala* *Pseudorca* *Sousa* *Stenella* *Steno* *Tursiops*
	Physeteridae	*Kogia*
	Ziphiidae	*Berardius* *Hyperoodon* *Mesoplodon* *Tasmacetus* *Ziphius*

* Notified in the *Australian Government Gazette* on 17 June 1975.
† Statutory Rules 1961, No. 65.

Printed by Authority by the Government Printer of Australia

13913/75—Recommended retail price 5c

Licence No. ....   Form 1.   Regulation 6(1).

COMMONWEALTH OF AUSTRALIA

Whaling Act 1960

LICENCE TO USE SHIP OR AIRCRAFT AS WHALE CATCHER

Subject to the conditions specified hereunder, ...

....

* Strike out whichever is inapplicable.

† Here insert the name of the ship.

‡ Here insert the nationality mark and registration mark of the aircraft.

of ....
the *owner   *ship†
    *charterer *aircraft‡
registered at .... is licensed
to use that *ship
        *aircraft as a whale catcher in the following area of waters:—

....

....

....

This licence is granted subject to the following conditions:—

§ Here insert the conditions subject to which the licence is granted.

1. § ....

....

....

....

Unless sooner cancelled in accordance with section 11 of the *Whaling Act 1960*, this licence remains in force until the .... day of .... 19.......

Dated this .... day of .... 19.........

....
Secretary

F32-272

Licence No.

FORM 2.   Regulation 6 (2.).

COMMONWEALTH OF AUSTRALIA.

Whaling Act 1960.

LICENCE TO USE A SHIP AS A FACTORY SHIP OR A FACTORY AS A LAND STATION.

Subject to the conditions specified hereunder,

of ....

the .... of the

* Strike out whichever is inapplicable.
*ship   registered at
*factory   situated at .... is licensed to use

that *ship / *factory as a ⁺land station / *factory ship for the .... of whales taken or killed in the following area of waters:—

....
....
....
....

This licence is granted subject to the following conditions:—

⁺ Here insert the conditions subject to which the licence is granted.

1. ⁺ ....
....
....
....
....

Unless sooner cancelled in accordance with section 11 of the *Whaling Act* 1960, this licence remains in force until the .... day of .... , 19 .

Dated this .... day of .... , 19 .

Secretary.

APPENDIX 8 : THE SPERM WHALE MODEL

The mathematical model used by the Scientific Committee of the International Whaling Commission in making assessments of sperm whale stocks has been progressively refined since its initial development in 1968 (see IWC/FAO, 1969; Allen, 1973; Allen and Kirkwood, 1977(a), (b) and (c)) and a brief overview is provided in Allen and Chapman (1977). A further modification to the details of the method for estimating population sizes from catch and effort data was introduced before the 1978 Scientific Committee meeting (Kirkwood, 1978). An alternative way of looking at the model which is also of interest is provided by Holt in Appendix 10 of his paper in the second volume of this report. Our intention in this appendix is not to cover all aspects of the model but rather to give the interested reader a broad understanding of what is involved.

The model

The model assumes that a percentage p (the pregnancy rate) of a stock of mature female sperm whales gives birth in any year. This percentage, which is usually in the range 20-25 per cent, allows for the period of lactation and resting between pregnancies. The offspring, half males and half females, are subject each year to a juvenile mortality rate M_j and this rate applies until they reach age t_j. From then on a (usually lower) natural mortality rate M applies. Females surviving to maturity at age t_{mf} aggregate socially into harem groups during the breeding season with an average of h females per breeding male. They continue to grow and become large enough to be taken by the whalers at age t_{rf} - the age at recruitment to the exploitable (female) population. Males become exploitable at age t_{rm}, which usually is at about the same age as they become sexually mature. However, males reaching sexual maturity do not participate in breeding unless they survive to reach social maturity at age t_{mm}, when they can seek to secure a harem.

To maintain the natural pregnancy rate for the female population the theoretical minimum number of socially mature males is one for every h mature females. It may, however, be necessary to have a reserve of socially mature males to stimulate the male with the harem to full breeding activity or if necessary to provide replacements during the breeding season. It is therefore assumed that there should be 1+r males for every h females (r is termed the reserve male ratio) and if the number of socially mature males falls below this, the model postulates that the pregnancy rate is lower than it could be by a factor equal to the ratio of the available number of socially mature males to the desired number.

Any additional socially mature males are surplus to breeding requirements and can be taken from the population. For a given female population, the maximum sustainable yield of males is thus obtained by keeping the ratio of socially mature males to mature females to (1+r):h, the minimum for a sustainable situation.[1]

It is also assumed that when the density of the population falls below the carrying capacity of the environment, a density dependent response in either births or natural deaths occurs to return the population towards its maximum. In the mathematical model provision is made for such adjustments in the pregnancy, juvenile mortality and natural mortality rates, and the female age at maturity. Currently a response through the pregnancy rate is adopted as the most likely process by which sperm whale populations adjust, and the other parameters are held constant. Details of the way this is incorporated in the model are set out later in this appendix.

The population components used in the model can thus be listed:
- juveniles of each sex - subject to a juvenile mortality rate different from that for older animals;
- young of each sex - post-juveniles, but sexually immature, subject to a mortality rate the same as that for older animals;

---

(1) Note that this assumes that the size composition of the catch is the same as that of the exploitable population. In fact whalers tend to select larger males when available, so that some adjustment for this imbalance would be desirable.

- unexploited mature females;
- exploitable mature females;
- exploitable socially immature males;
- socially mature males, with two elements: that number required for harems or as reserves, and the remainder which are surplus to breeding requirements and hence exploitable.

The number of animals in each component on a year by year basis is adjusted by allowing for births, deaths (both natural and due to whaling) and transitions from one component to the next.$^{(1)}$

It is necessary to establish the relative numbers in population components at some time in order to trace the changes through time. This is achieved by assuming that the population is stable before the commencement of exploitation, so that at that time the annual increase in each component was balanced by losses due to natural deaths or movements to the next component. It is unlikely however that the model would be internally consistent if independent estimates are used for all parameters. The juvenile mortality rate, which is one of the most difficult to measure directly, is the one used to achieve consistency. Its value is determined from the balance in the initial population between recruits and natural deaths for the mature female population.$^{(2)}$

Table 1 gives the values adopted in recent years for the parameters in applying the model to southern hemisphere sperm whale stocks.

---

(1) The number of individuals in a component surviving from one year to the next is basically assumed to follow a relationship of the form $N_{t+1} = N_t e^{-M}$.

(2) The balance equation is:
$$M_j = \frac{1}{t_j}\left\{\log_e \frac{p}{2(1-e^{-M_j})} - M(t_{mf}-t_j)\right\}$$
with the pregnancy rate being that for the stock before any exploitation.

Table 1

Population parameters for sperm whales in the southern hemisphere

			1976[1]	1977[2]	1978[3]
Average age at recruitment	male	(t_{rm})	20[4]	20[4]	20
	female	(t_{rf})	13	13	13
Average age at maturity (*social maturity)	male*	(t_{mm})	25	25	25
	female	(t_{mf})	8.5–10	10	10
Natural mortality rate	Juvenile (M_j) (for 2 years)		0.133	0.133	0.093
	Mature (M) (2 years +)		0.05	0.05	0.055
Mean number of mature females available to a school-master		(h)	10	10	10
Reserve male ratio		(r)	0.3	0.3	0.3
Pregnancy rate (p)	Unexploited		0.19	0.19	0.20
	Maximum		0.25	0.25	0.25
Density-dependent exponent		(n)	1.4	1.4	1.4

(1) Rep. Int. Whal. Commn 27 (1977), p.42
(2) Rep. Int. Whal. Commn 28 (1978), p.57
(3) Scientific Committee meeting 1978, Cambridge.
(4) For Division 9, t_{rm} = 15 was used.

The variation in pregnancy rates with density is assumed to operate through changes in the number of mature females. Figure 1 illustrates the kind of relationship used. The mathematical form chosen to represent this relationship[1] includes a parameter n (the density dependent exponent) which determines the convexity of the curve. If no density dependence is assumed (n=0) the pregnancy rate changes linearly with density and the maximum sustainable yield population level (MSYL) is at 50 per cent of the initial population size. Larger values of n generate a relationship such as that illustrated where the pregnancy rate responds to decreases in density at a rate more rapid than a linear rate. The rate of response

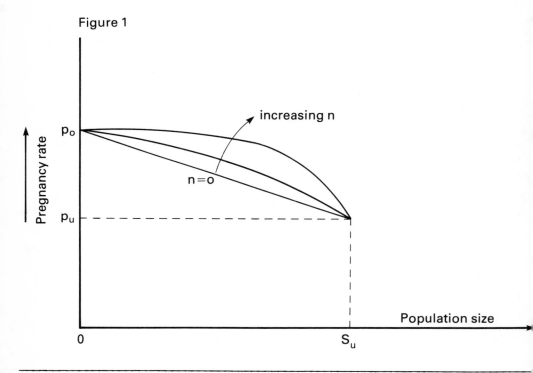

Figure 1

---

(1) The pregnancy rate for a stock S of mature females is given by the formula

$$p = p_o - (p_o - p_u)(S/S_u)^{1+n}$$

where p_u is the pregnancy rate when the population of mature females is at maximum density (and thus unexploited);

p_o is the upper limit for the pregnancy rate as zero population is approached;

S and S_u are respectively the current and unexploited numbers of mature females; and

n is the density dependent exponent

is greatest when the density first begins to decrease, and this rate continually decreases for further decreases in density, although still remaining greater than a linear rate. [1] The larger is the density dependent exponent, the closer is the MSYL to the initial population level.

The form of the density dependence relationship for sperm whales is at best an approximation, and the choice of the density dependent exponent is also largely arbitrary. Based on biological indications that the MSYL should be higher than 50 per cent of the initial population, scientists adopted a relationship corresponding to a MSYL for females of 60 per cent of the initial population (see Holt, Inquiry Report, Volume 2, pp. 32-36). This corresponds to a value of 1.4 for the density dependent exponent.

In summary then, the concept of maximum sustainable yield for sperm whales is different to that for baleen whales where the total population is assessed and catches are regulated to hold the population near a maximum sustainable yield population level of 60 per cent of the initial population. The sperm whale model separates the basis on which the yields for males and females are determined. The female stock is treated as though it has the capacity to produce the maximum sustainable yield at 60 per cent of its initial level. However, the yield of males is represented by the surplus beyond those males needed to ensure that harems are serviced effectively in the breeding season. Because of this interrelation, it is necessary to determine simultaneously the respective male and female population levels which produce the maximum combined yield.

It should also be noted that while the discussion so far has dealt with yields in terms of numbers of whales, and while the choice of the density dependent exponent is closely linked to the level at which maximum sustainable yield by number is expected to occur in the female population, this does not limit the use of the model to maximising yields by number.

---

(1) Note that an observed decline in the pregnancy rate for a particular stock of females could in principle result from either an increase in density or a shortage of socially mature males (unless the pregnancy rate falls below that for the unexploited population). However, changes of the first kind would generally be much smaller, and hence less likely to be detected, than those of the second.

Provided that it reflects the dynamics of the population, it can also be used to give estimates of male and female population levels which achieve other objectives, such as maximising yield by weight.

The following figures, which are broadly comparable although in some cases parameter values differ slightly, provide an illustration of the differences which can result from different objectives. In each case the populations referred to are the exploitable populations.

(a) Maximum yield of females alone by number is at 60 per cent of the initial female population.

(b) Maximum yield of males alone by number is with females at 100 per cent of the initial female population.

(c) Maximum yield by number of males and females combined is at 32 per cent and 76 per cent of the initial male and female populations respectively (derived from IWC, 1978, p.60 and p.88).

(d) Maximum yield by weight of males and females combined is at 35 per cent and 83 per cent of the initial male and female populations respectively (derived from IWC, 1978, p.60 and p.88).

Use of the model

Suppose that the values of the parameters are known. Then for any given number of mature females in the initial population, the number of animals in other initial population components can be calculated from the assumption of stability. The model can then provide the numbers each year in each population component (including into the future if desired) for the actual or any assumed series of annual catches for males and females. This process is incorporated in the SPDYN (sperm whale dynamic pool model) computer program (Allen and Kirkwood, 1977 (c)).

Since reliable direct estimates of sperm whale population sizes are not available, the problem can thus be looked at as one of finding the best estimate of the initial number of mature females. This is done using measures of comparative abundance of whales over time, and is based essentially on a principle which can best be illustrated by a simple example: if an abundance index is halved by a catch of 10,000 whales in one year, the population before the catch must have been 20,000 (ignoring here births and natural deaths).

Abundance indices have generally been derived from catch per unit effort (CPUE) data, although aircraft sightings have been used in the case of the Cheynes Beach operation in Division 5. At the June 1978 meeting of the Scientific Committee the raw CPUE data, recorded as male catch per catcher day worked, were modified for each stock and for each individual national operation to take account of changes in whaling efficiency (such as more efficient chasing resulting from the introduction of ASDIC) to derive time series for male catch per searching hour. These sets of refined CPUE data were seen as indices of male sperm whale abundance.

For each abundance index for a particular stock, that initial mature female population (or the corresponding initial population of exploitable males) is chosen which leads to changes in abundance that most closely approximate the actual changes expressed in the index when the history of actual total catches from the stock is taken into account. This is done using a maximum likelihood estimation procedure, together with the population dynamics in the SPDYN program, in the POPDYN computer program (Kirkwood, 1978).

Where there is more than one estimate for a stock of the number of mature females in the initial population, these estimates are combined. This has previously been done by simple averaging, but weighting to reflect the reliability of individual estimates may be undertaken in the future.

The SPDYN procedure is then applied for each stock using the best estimate of the initial mature female population and using actual catches, to give estimates of initial and current populations of exploitable males and females.

Another computer program, SPVAP (sperm whale variable population model), is used to determine the maximum sustainable yields for males and females and the maximum sustainable yield levels, given a set of parameter values (Allen and Kirkwood, 1977(b)). The program assumes an initial mature female population of 10,000 and produces a table of sustainable yields corresponding to various lower mature female populations. The maximum sustainable yield and maximum sustainable yield levels can be seen directly from this table and simple scaling then provides the desired results for any value of the initial mature female population. The program can also produce yield estimates in terms of weight using relationships between age, length and weight. It thus provides the basis for calculating the population levels and resultant yields corresponding to a wide range of possible management objectives once population parameters and the initial population size are determined.

The classification of stocks and the calculation of catch limits now follow from a comparison for each sex of the current population level with the maximum sustainable yield level in accordance with the usual rules as outlined in Chapter 3 of the report.

APPENDIX 9: POSSIBLE OBJECTIVES FOR MANAGEMENT OF MARINE MAMMALS

The material in this appendix is Table 10.1 in FAO/ACMRR (1978).

The three headings within the list of objectives are used as points of reference rather than as clear boundaries between the named objectives; each objective overlaps, and in some cases can be seen to conflict, with others under the same heading and under different headings. All objectives stated must be considered in relation to both long-term sustainable benefits and short-term or immediate benefits.

Socio-economically oriented objectives

1. Providing commodity yield (including food, industrial products and so on),
    (a) from marine mammals
    (b) from competitors of marine mammals (e.g. fish at high trophic levels)
    (c) from food species of marine mammals (e.g. krill).

2. Providing recreation and tourism,
    (a) oriented toward hunting and fishing for sport
    (b) oriented toward nature observation (e.g. whale watching).

3. Providing employment.

4. Providing cash income.

5. Providing for cultural diversity (e.g. survival of traditional and subsistence economies).

6. Providing for distribution of benefits to all levels of society,
    (a) nationally
    (b) internationally.

7. Providing for scientific uses and increases of knowledge.

8. Providing education benefits.

9. Providing for human health.

10. Providing for domestication (e.g. as sources of food and other commodities and as work animals).

Ecologically oriented objectives

11. Maintaining ecosystem diversity.

12. Maintaining ecosystem stability.

13. Maintaining gene pools, distribution of species and varied environments.

14. Maintaining ability of populations to survive fluctuating environmental conditions.

Ethically oriented objectives

15. Providing minimum stress for marine mammals.

16. Increasing survival chances of marine mammals (including not killing).

17. Particularly respecting the life of cetaceans because of their intelligence, friendliness and lack of aggressive behaviour toward man.

18. Avoiding inhumane or cruel practices involving marine mammals.

19. Maintaining the options for future generations of human beings.

20. Not killing animals at all.

APPENDIX 10: WHALING STATISTICS

Table 1: Current and initial sizes of stocks of the great whales (including minke whales), their classification and catch limits for the 1978-79 and 1979 seasons

Table 2: Annual southern hemisphere sperm whale stock classifications, catch limits and catches by sex, management region and country for 1971-72 and 1972 seasons to 1978-79 and 1979 seasons

TABLE 1: CURRENT AND INITIAL SIZES OF STOCKS OF THE GREAT WHALES (INCLUDING MINKE WHALES), THEIR CLASSIFICATION AND CATCH LIMITS FOR THE 1978-79 AND 1979 SEASONS

Species	Stock or area	Stock number (thousands)* Current (year)	Initial (year)	% of initial	% above or below MSYL	Classification	Catch limit	Source	Notes
SOUTHERN HEMISPHERE									
Blue	All stocks	5	180	3%	-95%	PS	0	(1)	Possibly some local concentrations beginning to rebuild
Pygmy blue	All stocks	5	10	50%	-16%	PS	0	(1)	Possibly close to MSY level
Humpback	All stocks	3*	100	3%	?	PS	0	(1)	
Right	All stocks	3.8-4.8*	100+?	4%	?	PS	0	(1), (2)	Increases observed off South Africa
Fin	Area I	3.1	16.8	18%	-69%	PS	0	(2),(4)	
	Area II	19.4	124	16%	-72%	PS	0	(4)	
	Area III	38.8	152	26%	-55%	PS	0	(4)	
	Area IV	8.4	60	10%	-75%	PS	0	(4)	
	Area V	3.1	28	11%	-81%	PS	0	(4)	
	Area VI	12.4	24	52%	-11%	PS	0	(4)	
Sei	Area II	1.7-3.0	5.3-7.4 ('51)	30-40%	-50 to -33%	PS	0	(4)	Indications of increasing population prior to exploitation complicate use of initial population size as a reference level
	Area III	5.1	11.4 ('65-7)	45%	-25%	PS	0	(4)	
	Area IV	1.4-3.0	20.-21.3 ('60)	7-14%	-86 to -77%	PS	0	(4)	
	Area V	6.1	18.0 ('60)	27-31%	-55 to -48%	PS	0	(4)	
	Area VI	3.8-4.4	13.9-14.3 ('60)	24% of '60	-60%	PS	0	(4)	
Bryde's	Indian Ocean	10	?	?	?		0	(4)	
	South Pacific (west)	?	?	?	?	IMS		(4)	120 take by scientific permit
	South Pacific (east)	?	?	?	?			(4)	Approx. 400 taken annually by Peru
Minke	Area I	7.8	9.4		Population expanding prior to exploitation so classification not attempted; catch limits based on replacement yields		738^a	(4)	a:10% allowance included; area catches must not exceed limit shown. However, total for all areas must not exceed 6221
	Area II	22.6	21.7				1272^a	(4)	
	Area III	51.2	42.6 ('71-2)				2510^a	(4)	
	Area IV	16.4	25.4				139^a	(4)	
	Area V	6.1	6.6				563^a	(4)	
	Area VI	5.7	4.3				371^a	(4)	
						total	6221		
Sperm (male)	Division 1	4.6	12.4	37%	+15%	SMS	273	(4)	**:limits represent 1977-78 and 1978 season limits or are adjusted down by 10% or 25%; no 10% allowance was incorporated in Divisional limits
	Division 2	16.4	33.0	50%	+53%	DMS	808	(4)	
	Division 3	13.7	38.4	36%	+10%	SMS	847	(4)	
	Division 4	12.1	23.1	55%^{b}	+61%	SMS	566	(4)	
	Division 5	4.7^b	18.2^b ('46)	26%^{c}	-19%^{b}	SMS	402	(4),b:(5)	
	Division 6	5.1	11.3	40%^{c}	+25%^{c}	IMS	276	(4),c:(6)	
	Division 7	4.9	15.9	31%	-4%	SMS	176	(4)	
	Division 8	33.6	37.7	89%	+175%	IMS	874	(4)	
	Division 9	12.8	36.2	35%	-14%	PS	0	(4)	
Sperm (female)	Division 1	17.8	20.3	88%	-16%	SMS	91	(4)	**as above
	Division 2	50.8	54.1	94%	+24%	SMS	241	(4)	
	Division 3	52.6	63.0	84%	+10%	SMS	281	(4)	
	Division 4	21.6^d	37.9^d	60%^{d}	-25%^{d}	IMS	0	(4),d:(5)	
	Division 5	24.3	26.7 ('46)	91%^{e}	-20%^{e}	IMS	159	(4),e:(6)	
	Division 6	17.3	18.5	92%	+6%	SMS	83	(4)	
	Division 7	21.0	26.5	83%	-6%	SMS	98	(4)	
	Division 8	57.4	58.5	98%	+29%	IMS	261	(4)	
	Division 9	25.7	46.4	55%	-27%	PS	0	(4)	
NORTH PACIFIC									
Blue	All stocks	1.6	8.9	33%	-45%	PS	0	(1)	
Humpback	All stocks	1.4*	?	?	?	PS	0	(1)	
Right	All stocks	0.15	?	?	?	PS	0	(1)	
Gray	Eastern stock	11f,15f,**	15f,**	?	?	SMS	178^h	f:(4),g:(7)	h: catch for aboriginal use only
	Western stock	>20*	?	?	?	PS	0	(7)	Stock possibly extinct

266

TABLE 1 CONTINUED

Region	Species	Stock	MSYL	Current (year)	% of MSYL	Current %	Status	Catch	Notes
NORTH PACIFIC	Fin	All stocks	18 ('76)	44	41%	-33%	PS	0	(2)
	Sei	All stocks	9 ('75)	43	21%	-60%	PS	0	(8)
	Bryde's	Eastern stock	10*	21 ('48)	?	?	IMS	0[d]	(4)
		Western stock	15.6		7%	+23%	IMS	454	(4)
	Minke	Okhotsk Sea stock	?	?	?	?	SMS		(4)
		Sea of Japan stock	?	?	?	?	SMS	460	(4)
		Remainder	?	?	?	?	IMS	0[j]	(4)
	Sperm (male)	Western Division	52.7	123.4 (1970)	43%	-7%	SMS	k	(4)
		Eastern Division	111.4	142.7 (1970)	78%	+70%	IMS	k	(4)
	Sperm (female)	Western Division	110.2	145.6 (1910)	76%	-6%	SMS	k	(4)
		Eastern Division	162.6	158.3 (1910)	97%	+20%	IMS	k	(4)
ARCTIC	Bowhead	Bering Sea	2.26*	18* (1850)	13%	-78%	PS	18(or 27 hit)	(3),(4)
		Spitzbergen	>0?	25*	?	?	PS	0	(3)
		Davis Strait	0.6*	6*	10%	-83%	PS	0	(3)
		Hudson Bay	0.1*	0.7*	15%	?	PS	0	(3)
		Okhotsk Sea	>0?	6.5* (1679)	?	?	PS	0	(3)
	Blue	All stocks	0.1	1.1-1.5	7-10%	?	PS	0	(1)
	Humpback	Western stock	1.0[l]-2.3[m]*	1-1.5[l]	?	?	PS	0	1:(1) m:(4)
		Eastern stock	>0?	?	?	?	PS	0	(4)
	Right	Western stock	.07	?	?	?	PS	0	(3)
		Eastern stock	>0?	?	?	?	PS	0	(3)
NORTH ATLANTIC	Fin	Nova Scotia	.43 ('76)	1.2	36%	-39%	PS	0	(2)
		Newfoundland-Labrador	?	?	?	?	IMS	90	(2)
		West Greenland	1.9 ('76)	2.4	79%	+33%	SMS (Pr.)	15	(4)
		East Greenland-Iceland	?	?	?	?	SMS	304 max.	(4)
		North Norway	8.1	?	?	?	SMS (Pr.)	61	(3)
		West Norway (Faroes)	1.1-1.6 ('63)	2.7-6.0 (pre '46)	?	?	PS	0	(2)
		Spain-Portugal-British Isles	?	>10.6 (pre '21)	?	?	SMS (Pr.)	-	(2)
	Sei	Nova Scotia	0.9-2.2 ('76)	?	?	?	PS	0	(2)
		Iceland-Denmark Strait	?	?	?	?	SMS	84	(2)
	Bryde's	All stocks	?	?	?	?	IMS	0	(4)
	Minke	Canadian East Coast	?	?	?	?	SMS	48	(2)
		West Greenland	?	?	?	?	SMS (Pr.)	394	(2)
		East Greenland-Iceland-Jan Mayen	?	?	?	?	SMS	320	(4)
		Svalbard-Norway-British Isles	50.6	?	?	?	SMS	1790	(4)
	Sperm (males and females)	All stocks	22 ('71)*	?	?	?	SMS	665	(9)
	Bottlenose	All stocks	?	?	?	?	PS (Pr.)	0	(3)

* Stock numbers relate to exploitable populations except where asterisks shown - these denote estimates of total population number.

MSYL - maximum sustainable yield level
PS - protected stock
IMS - initial management stock
SMS - sustained management stock
SMS(Pr.) - provisional sustained management stock
PS (Pr.) - provisional protected stock

Notes:
- i: pending satisfactory estimate of stock
- Mean of 1965-76 catch
- Catching effort should not increase; exploited by non-IWC nations
- j: pending satisfactory estimate of stock
- k: catch limits deferred pending re-examination of stock identity and assessments
- Eskimo catch only
- Rare catches by Spain
- Extremely rare catches by Portugal
- Subsistence catch not to exceed 15 fin and humpback combined
- Total catch 1977-82 not to exceed 1524
- Limit represents 1948-71 average
- Protected in absence of new data
- Catches should not increase; exploited by non-IWC nations
- Protected in absence of new data
- Limit represents 5 year average
- Pending satisfactory estimate of stock
- Limit represents average of recent catches
- Average of '68 - '77 catch
- Average of '61 - '75 catch
- 10 year average catches
- Average of '69 - '73 catch; exploited mainly by non-IWC nations
- Pending information for classification

References:
(1) - Gambell (1976)
(2) - Scientific Cttee Rpt,1976
(3) - Scientific Cttee Rpt,1977
(4) - Scientific Cttee Rpt, 1978
(5) - Kirkwood, et. al. (1978)
(6) - Kirkwood (pers. comm.)
(7) - Wolman and Rice (1978)
(8) - Scientific Cttee Rpt, 1974
(9) - Scientific Cttee Rpt, 1971

TABLE 2: ANNUAL SOUTHERN HEMISPHERE SPERM WHALE STOCK CLASSIFICATIONS, CATCH LIMITS AND CATCHES BY SEX, MANAGEMENT PERIOD AND COUNTRY FOR 1971-72 AND 1972 SEASONS TO 1978-79 AND 1979 SEASONS

Season	Country	Male quota and (catch)	Female quota and (catch)	Division 1 catch limit male:female (catch)	Division 2 catch limit male:female (catch)	Division 3 catch limit male:female (catch)	Division 4 catch limit male:female (catch)	Division 5 catch limit male:female (catch)	Division 6 catch limit male:female (catch)	Division 7 catch limit male:female (catch)
1978-79 and 1979 Season[a]	USSR (NA)	3068 (NA)	840 (NA)	}a	}a	}a	}a	}a	}a	}a
	Japan (NA)	736 (NA)	200 (NA)							
	Australia (NA)	402 (NA)	159 (NA)					402:159 (NA)		
	Brazil (NA)	16 (NA)	15 (NA)							
IWC Stock Classification				IMS:IMS	SMS:SMS	PS:PS	SMS:SMS	SMS:IMS	IMS:PS	IMS:IMS	SMS:SMS
Total Catch Limit[b] 5436		-	-	874:261	273:91	0:0	808:241	847:281	566:0	402:159	276:83	176:98
(Catch)		4222 (NA)	1214 (NA)	(NA)	(NA)	(NA)[c]	(NA)	(NA)	(NA)	(NA)	(NA)	(NA)
1977-78 and 1978 Season	USSR[p] (4241)	3213 (NA)	950 (NA)	}e	}e	}e	}e	}e	}e	}e
	Japan (296)	772 (NA)	226 (NA)	(783:291)	(325:94)	(0:0)	(731:292)	(753:283)	(531:0)	(44:0)	(80:0)	(229:101)
	Australia (NA)	536 (NA)	177 (NA)					536:177 (NA):(NA)		
	Brazil (NA)	17 (NA)	17 (NA)		17:17 (NA)(NA)					
IWC Stock Classification				IMS:IMS	SMS:SMS	PS:PS	SMS:SMS	SMS:SMS	IMS:PS	SMS:IMS	IMS:IMS	SMS:SMS
Total Catch Limit[d] 5908		4538[d]	1370[d]	961:319	333:111	0:0	889:295	1035:343	623:0	590:195	304:101	257:143
(Catch)	(NA)	(NA)	(NA)	(763:299)	(NA)	(0:0)[c]	(731:292)	(753:283)	(531:0)	(NA)	(80:0)	(229:101)
1976-77 and 1977 Season	USSR (3841)	3134 (3134)	707 (707)	}e	}e	}e	}e	}e	}e	}e
	Japan (234)	238 (174)	63 (60)	(745:208)	(296:62)	(0:0)	(785:194)	(632:183)	(1590:0)	(47:23)	(213:3)	(0:94)
	Australia (624)	508 (508)	116 (116)					508:116 (508:116)		
		1 whale lost at sea not included								
	Brazil (25)	14 (13)	11 (12)		14:11 (13:12)					
IWC Stock Classification				IMS:IMS	SMS:SMS	PS:PS	IMS:IMS	SMS:SMS	IMS:PS	SMS:IMS	IMS:SMS	PS:SMS
Total Catch Limit[d] 4791		-	-	909:209	316:73	0:0	840:194	783:224	590:0	559:128	287:66	0:94
(Catch)[c]	(4724)	3890[d] (3829)	897[d] (895)	(745:208)	(309:74)	(0:0)[c]	(785:194)	(632:183)	(590:0)	(555:139)	(213:3)	(0:94)
1975-76 and 1976 Season	USSR[h]	3658 (3658)	2796 (2796)	1219:785 (731:839)[k]	1619:1570[f] (1245:989)[k]	96:0[g]	← 388:424[g]	729:400[f] (406:0)[g] (417:371)[k]		340:217 (284:73)[g]	0:262 (0:0)[g]	399:320 (455:328)[k]
	Japan[h]	878 (364)	665 (228)	293:187	387:373[f]		← 175:95[f] →		82:52	0:62	96:76
				g	g		g	g	g	g	g
	Sth Africa	658 (0)	873 (0)					658:873 (0:0)		
	Australia	658 (658)	487 (337)					658:487 (658:337)		
	Brazil	18 (4)	49 (5)		18:49 (4:5)					
IWC Stock Classification		-	-	IMS:IMS	SMS:IMS	SMS:SMS	IMS:IMS	SMS:SMS	SMS:SMS	IMS:IMS	PS:SMS	SMS:SMS

Season	Country	Male quota and (catch)	Female quota and (catch)	Area VI/Area I catch limit male female	Area II/Area III catch limit male female	Area IV/Area V catch limit male female
1974-75 and 1975 Season	USSR	4985 (4985)	2871 (2871)	2710:969 (1640:442)g (752:68)g	1212:1286)g 1063:892 (213:779)g	1212:1010 (1320:1283)g (391:145)g (291:240)
	Japan	1196 (543)	683 (531)	650:231	255:212	
	Sth Africa	897 (868)	896 (810)		897:896 (868:810)	
	Australia	897 (692)	500 (480)			897:500 (692:480)
	Brazil	25 (9)	50 (45)		25:50 (9:45)	
Total Catch Limit[h] 13,000 (Catch)[c] (11,838)		8000[b] (7097)	5000[b] (4737)	3822:1500 (2392:510)e	2548:2563 (2302:2319)	2730:2188 (2403:1908)
1973-74 and 1974 Season	USSR (7900)	5000 (5165)[j]	2900 (3149)[j]	??	??	??
	Japan (415)	1200 (j)	690 (j)			
	Sth Africa	900 (882)	905 (899)		900:905 (882:899)	
	Australia	900 (630)	505 (449)			900:505 (630:449)
Total Catch Limit[l,m] 13,000 (Catch)[l,m] (11,177)		8000[k] (6677)	5000[k] (4497)	k3200:1100 (NA) (1)	k1900:1800 (NA) (m)	k2900:2100 (NA)
1972-73 and 1973 Seasons	USSR	5000 (4998)	2900 (2900)	??	??	??
	Japan	1200 (445)	690 (398)			
	Sth Africa	900 (860)	905 (746)			
	Australia	900 (694)	505 (287)			
Total Catch Limit 13,000 (Catch)[n,o] (11,316)		8000 (6986)	5000 (4330)			
1971-72 and 1972 Seasons		NO CATCH LIMITS FOR SPERM WHALES IN THE SOUTHERN HEMISPHERE WERE SET BY THE INTERNATIONAL WHALING COMMISSION FOR THE 1971-72 SEASONS OR PREVIOUS SEASONS				

(a) Details of USSR and Japan Divisional allocations not provided; catch data not yet available.
(b) No 10 per cent allowance between Divisions to be allowed for this season.
(c) Peru and Chile take males and females from Division 9; a catch of 841 whales was reported in 1975 but later figures have not been provided.
(d) Ten per cent allowance between Divisions has been included in Divisional catch limits shown; catch in any Division shall not exceed limits shown but sum of all Divisional catches must not exceed total catch limit.
(e) Details of USSR and Japan Divisional allocations not provided; combined catches shown in brackets.
(f) 1975-76 and 1976 catch limits were set for six Southern Hemisphere regions based on sperm whale management Divisions but grouping Divisions 3 and 4 and Divisions 9, 1 and 2.
(g) Combined USSR and Japan catches shown; separate catch data by regions not available.
(h) An allowance between regions has been included in regional catch limits shown; catch in any region shall not exceed limits shown but sum of all regions must not exceed total catch limit.
(i) 1974-75 and 1975 catch limits were set for three regions based on combinations of Antarctic baleen whale management areas.
(j) Combined total pelagic catch shown; no national or area breakdown published.
(k) USSR and Japan objected to Division 9; this catch limits by areas so the area limits were not binding on them.
(l) Peru and Chile took 1286 and 130 whales respectively from Area I; these catches not included.
(m) Brazil took 29 whales from Area II; this catch not included.
(n) Peru and Chile took 1497 and 232 whales respectively from Area I; these catches not included.
(o) Brazil took 75 whales from Area II; this catch not included.
(p) In a previous season Japan and USSR re-distributed their allocation privately; this may explain the discrepancy between catch and the allocation shown.

ABBREVIATIONS
NA - data not yet available
SMS - Sustained Management Stock
IMS - Initial Management Stock
PS - Protected Stock

(Source: International Whaling Statistics; Schedule to the International Convention for the Regulation of Whaling; Reports of the International Whaling Commission; Commonwealth Department of Primary Industry)

APPENDIX 11: AN ASSESSMENT OF THE SPERM WHALE STOCK
SUBJECT TO WESTERN AUSTRALIAN CATCHING

(A submission to the Inquiry by G.P. Kirkwood, K.R. Allen and J.L. Bannister

The stock of sperm whales that has been exploited by Australian catchers is recognised by the International Whaling Commission (IWC) Scientific Committee as falling into Division 5 of the southern hemisphere, an area south of the equator lying between longitudes 90°E and 130°E. At its recent meeting in Cambridge, the Scientific Committee agreed on an appropria data base and analytic procedures for the assessment of southern hemisphere sperm whale stocks, but time did not permit these assessments to be carried out. In this submission, an assessment of the Division 5 stock has been carried out, along the lines recommended by the Scientific Committee.

Recently detailed data from Australian whaling operations at Albany have been re-examined, and revised series of indices of relative abundance based on aircraft sightings data have been obtained. Two series were extracted; the first represents the number of exploitable males seen per flying day in a restricted area that has been searched since operations commenced, while the second represents total numbers of exploitable males seen per flying day. In 1967 the original float plane spotter aircraft was replaced by a twin engined aircraft. Recognising the increase in efficiency of this new aircraft, only data from 1968 onwards were used in the assessment. During the 1970s the area searched expanded, inevitably reducing to some extent searching effort in the more restricted original search area. Thus it is likely that declines in relative abundance seen in the restricted series will overestimate the true decline. On the other hand, an increase in the total area searched will result in more whales being seen, and the decline in relative abundance in the second series is likely to underestimate the true decline. Consequently estimates of initial and current stock sizes based on the restricted area sightings are likely to be underestimates, while those based on the total area sightings are likely to be overestimates, with the best estimates lying between the two. The sightings data are discussed in greater detail in the appended paper.[1]

In addition to the Australian sightings data, catch per unit effort data are available for whaling operations in Division 5 by other nations. At Cambridge, the Scientific Committee adopted the searching hour as a unit of effort, rather than the catcher day used in the past. In this assessment, two series of catches per searching hour were used, obtained from Japanese operations outside the baleen season (OBS) and Soviet operations north of 40° south (N40S) in Division 5. Other shorter series of data which were utilised in previous assessments could not be used on this occasion as they required additional data for the same countries from other Divisions, and these data are not presently available.

Using the model structure, parameter values and procedures adopted by the Scientific Committee at Cambridge, the following estimates of the exploitable female (aged 13 and over) and exploitable male (aged 20 and over) population sizes at the beginning of 1947 and 1979 were obtained.

Estimated population sizes in thousands for each data set

Year	Australian sightings total area		Australian sightings restricted area		Japanese OBS		Soviet N40S	
	Female	Male	Female	Male	Female	Male	Female	Male
1947	27.0	18.4	25.7	17.5	21.5	14.6	32.6	22.2
1979	24.5	4.9	23.2	4.0	18.7	1.2	30.1	8.7

These lead to average 1947 exploited population estimates of 26,700 for females and 18,200 for males. Using these estimates as starting values, the estimated population size each year was obtained by simulation using the actual catches and the sperm whale model. The results are shown in the following table.

---

(1) The appendix to this submission has not been printed here. It is 'Revised abundance indices for sperm whales off Albany, Western Australia' by G.P. Kirkwood and J.L. Bannister, which was submitted to the 1978 annual meeting of the International Whaling Commission's Scientific Committee (SC/30/Doc.22). For a description of the modelling procedure on which this assessment is based, see Chapter 4 and Appendix 8.

Estimated population sizes in thousands

Year	Exploitable females	Exploitable males
1947	26.7	18.2
1979	24.3	4.7

These analyses show that on the basis of the model now used by the Scientific Committee, the exploitable male and female sperm whales in Division 5 are respectively at 26 per cent and 91 per cent of their 1947 levels, levels which may be taken as having been close to stability. Applying the rules of the 'New Management Procedure' of the IWC, the male stock would be classified as a Protection Stock, with a zero catch limit, while the females would be classified as an Initial Management Stock with a catch limit of 131 for the 1978-79 pelagic season and the 1979 coastal season.

However it seems very likely that pregnancy rates in this stock have declined quite seriously in the last few years. At Cambridge, the data on observed percentage of pregnancies amongst mature females, provided by J.L. Bannister, were examined by the Scientific Committee and the percentages actually pregnant were estimated from the percentage pregnant and doubtful using the sizes of the corpora lutea. These data are presented below:

Estimated percentages of pregnant mature females taken

Year	% pregnant
1964	37.6
1965	29.3
1973	28.2
1974	22.2
1976	12.5
1977	22.6

Analysis of these data showed a statistically significant decline in percentage pregnant. It should be noted that observed percentages pregnant among captured mature females will almost certainly overestimate the true population pregnancy rates due to protection offered to lactating females. Although the earlier observations were collected by experienced biologists, while later observations were collected by inspectors without scientific training, it is felt that any bias due to the differing observers is small, and that the decline observed in these data should be taken seriously.

Examination of the results of the simulation run shows that there would have been a continuous drop in true pregnancy rate from 0.201 in 1966 to 0.067 in 1977. This drop results from the fact that, in the model, which uses actual catches, the ratio of socially mature males to mature females has dropped below the critical value needed to maintain the normal pregnancy rate for a given level of the female population. The decline in simulated pregnancy rate is actually greater than that in the observed data. This suggests that, if anything, either the model is somewhat conservative, or the initial population size assumed is too small. Nevertheless the good qualitative agreement between the observed pregnancy data and the model results suggests that some confidence may be placed in the model and extrapolations into the future based on the model as representing the likely course of the real populations.

This being so, the exploitable populations of both sexes can be expected to undergo a decline for a time. This will continue until the pregnancy rate has risen as a result of the sex ratio rising again above the critical level and the resulting increased number of recruits has entered the exploitable population. It should be noted that, assuming equal numbers of both sexes are born and equal mortality rates for each sex, one male is recruited to the socially mature stock for about every two mature female recruits, whereas only one male is required for about every seven mature females to maintain the normal pregnancy rate. Normal recruitment will therefore rapidly restore the conditions required to give normal pregnancy rates, in the absence of exploitation.

However the effect of the low pregnancy rate in recent years is to reduce recruitment to the mature and exploitable classes at the present time, and this will continue for some years, whether any further exploitation occurs or not. The simulation run with the model shows that even without exploitation, the number of exploitable females will decline for some years, and will actually fall into the Protection Stock category in about 1989, and will then take another 35 to 40 years to reach the MSY level again. At its lowest level the exploitable female stock will be 65 per cent of its initial size. In these circumstances, there seems good reason to cease taking females from this stock immediately, in spite of the fact that under current IWC rules a quota could be taken for some ten years as female numbers reduce from their current high level. Immediate cessation would have the effect of minimising the extent of the decline, although some reduction in stock size will inevitably occur.

This situation, in Division 5, seems to parallel quite closely that existing in the western North Pacific. In this case also, the Scientific Committee observed at its recent meeting in Cambridge that the data indicated a current decline in pregnancy rate which is in agreement with the effects of changes in sex ratio as shown by the simulations. The Scientific Committee, while pointing out that under the rules a catch was allowable, did 'not wish to recommend catch limits for either sex without drawing attention to the fact that...the ratio of socially mature males to sexually mature females fell below that determined by a reserve male ratio of 0.3...the analysis indicated that even if no catch is taken the stock will decline.'

Thus the present rules of the IWC, which define desirable stock levels and quotas in terms of stable situations, are inadequate in dealing with the dynamic situation which exists when a stock has already been reduced. This is one of the problems which will be examined in the review of the management procedure which the IWC initiated at its recent meeting.

When the females have returned to the MSY level, the taking of a sustainable yield would be permitted under the present rules, and this could be taken on a continuing basis without reduction of the stocks. The males however present a different situation. After relatively few years from the present, the number of socially mature males in the population would be expected to be above that required for reproduction, and some harvest could then be taken on a continuing basis. This harvest, if properly determined, could be continued indefinitely. In the dynamic situation which would exist while the female population reduced and then expanded, the factors affecting the safe male harvest would be complex and would require careful monitoring. They would include not only the size and recruitment rate of the male population, but also the size of the mature female population and the rate at which it was growing. Given the necessary knowledge, the general principles incorporated in the present management procedure would maintain the population of both sexes at or above the MSY level once the mature female population had been brought to it.

It may be noted that other simulations show that if this current procedure had been correctly applied from the beginning (i.e. with quotas of 464 males and 131 females) the populations would have stabilised well above MSY level. The catches of males which have been taken for a number of years would now appear to have been well above the quota level of 90 per cent of MSY.

APPENDIX 12:   AUSTRALIAN HUMPBACK WHALING

(A review prepared for the Inquiry by R.G. Chittleborough)

Introduction

Australian shore-based humpback whaling began in 1949 with one station, expanded to six stations by 1956 and then ground to a halt in 1963.

Regulation of this whaling was controlled by the Commonwealth Department of Commerce and Agriculture (later the Department of Primary Industry) which represented Australia at the International Whaling Commission. The CSIRO Fisheries Division (later the Division of Fisheries and Oceanography) undertook responsibility for research upon local stocks of humpback whales, and established links with the Scientific Committee of the International Whaling Commission. I commenced the Division's research upon this species in 1951, initially based on the west coast but in time collected research data from each of the five mainland stations and the station at Norfolk Island. In addition to material obtained from the commercial catch, I carried out whale marking on both the western and eastern coasts of Australia and made further observations from spotting aircraft. During the height of the research program upon this species, two scientists and three technical officers constituted the core of the research team with considerable involvement from other members of the Division and assistance from whaling inspectors.

At the opening of this period of Australian whaling, the International Whaling Commission was struggling through its early years which were dominated by industry and politics. The Scientific Committee contained some very able scientists with a great deal of knowledge of whales and a very real concern for their conservation. However in those days conservation did not have the momentum which is more evident today, so that the major whaling companies and the objective of maximising profits came to the fore during the meetings of the International Whaling Commission.

Regulation of baleen whales in the Antarctic was by an overall quota. This was expressed in blue whale units; in this one blue whale was equal to two fin whales or two and a half humpback whales or six sei whales. These ratios were based on equivalent oil yields, the emphasis again being on stabilising oil production rather than on management of stocks. At that time there was no attempt to manage stocks of individual species.

By setting an overall quota for baleen whales in the Antarctic the International Whaling Commission encouraged over-capitalisation and inefficient operations of factory fleets. From the opening date of the summer whaling season each fleet attempted to catch whales as quickly as possible in order to gain the highest proportion of the overall quota. This meant that there was very little selection of the larger whales, the only selection being to take those above the minimum length set for each species.

The humpback whale had been protected from pelagic whaling in the Antarctic during the period from 1939 to 1949. Recognising that this species was not particularly abundant, the International Whaling Commission set a quota of 1250 humpback whales as the Antarctic limit for that species to be taken by pelagic fleets in the summer of 1949-50. This then represented 500 blue whale units out of the overall Antarctic quota for that season of 16,000 blue whale units. Although this was the first time that a quota had been set for an individual species, it took no cognisance of the fact that the southern stocks of humpback whales were divided into six fairly separate populations. The modest quota of 1250 could be taken from whichever population the whaling fleets happened to be near; in fact most of this quota was taken to the south of Australia. The Antarctic quota of 1250 humpback whales was set without taking into account catches being taken or being planned in temperate waters at the northern end of the migration routes of these populations of humpback whales. Thus there was no real co-ordination of the control of humpback whaling.

In order to apply the limit of 1250 humpback whales, each factory ship was required to radio the humpback whale catch at the end of each day so that the closing date for humpback whaling could be announced when the target was being approached. However, under this scheme the catch of humpback whales exceeded the quota so in later years a fixed humpback whaling season, usually of only four days, was applied. At the end of that period the whaling fleets confined their attention to other species such as fin, blue and sei whales. There were infractions of this regulation on occasions, as will be mentioned later.

When Australian humpback whaling began, the Australian Government set quotas for each station. These quotas appear to have been based rather more on the capacity of the shore-based factory than upon the state of the whale population. The Australian Government also applied a legal minimum length of 35 ft, and prohibited the taking of females which were accompanied by a calf. Australian whaling operated upon two separate populations of the humpback whale. The population which came to the western coast of Australia each winter to breed, migrated to Antarctic waters between 70°E and 130°E for the summer feeding season. This was known as the Group IV population. Further eastward there was the Group V population of humpback whales feeding each summer between 130°E and 170°W. When this population migrated north each autumn, some moved along the eastern coast of Australia and others passed New Zealand and moved amongst the southwest Pacific islands including Norfolk Island. While migrating along our coast the humpback whales remained close to the shore, most being within ten miles of the coast. This made them particularly vulnerable to shore-based operations.

From the large amount of data collected throughout this phase of Australian whaling, we have built up a fairly clear picture of the impact of whaling upon these two populations of humpback whales. With the benefit of hindsight one can consider each population in turn and trace the effect of the catches upon population size and structure.

The Group IV (Western Australian) population: 70°E-130°E

The original size of this population before any hunting had taken place was in the region of 12,000-17,000 individuals. At this level, birth rate was balanced by natural deaths so that the population was close to its maximum at the environmental pressures operating at that time. At a level of 10,000 individuals, this population would have had its maximum rate of increase, i.e. the difference between birth rate and natural death rate would have been maximal. The net increase would then have been approximately 390 whales per year. This means that if the annual catch from the population then exceeds 390 humpback whales the population will decline.

Had it been possible to obtain this information before the commencement of whaling there would have been prospects for far better management. However these data were obtained by carefully recording the changes in population structure during the course of whaling so that by the time the necessary information was at hand, the permissible level of hunting was well below 390 individuals per year. By introducing these results here, a review of actual killing rates within this population can be put into perspective.

Table 1 lists the annual catches known to have been taken from this population during the present century. In the first phase (1912-16) humpback whales were hunted from two stations on the Australian coast, one near Albany and the other at Pt Cloates. These stations were owned and manned by Norwegians. While catch rates declined to some extent from year to year, this initial phase of fishing on a previously unfished stock only reduced it to about the size at which the replacement rate was maximised. Therefore by the time that the next phase opened the population had virtually recovered to its original size. It is of interest to note that at the end of this first era, the station near Albany turned to sperm whaling; another station at Albany made a similar change from humpback whales to sperm whales nearly half a century later.

During the second phase of humpback whaling (1925-28) the Norwegians re-opened the station at Pt Cloates. In terms of population size at the opening of this phase (approximately 15,000 whales), the stock had not been depleted when a dispute with the Australian Government over tariffs closed that station in 1928.

The third phase of hunting in this population (1935-39) caused severe depletion. Hunting took place both on the western coast of Australia (two factory ships operating from Shark Bay) and on the Antarctic feeding grounds. Catch rates fell markedly and the size composition of the population also declined. The population then had ten years to recover before whaling was resumed on a modest scale in 1949.

At the beginning of the recent era of whaling from this stock the total population contained approximately 10,000 individuals. With hindsight we now know that at this level the taking of more than 390 humpback whales per year would certainly cause depletion. Table 1 shows that for thirteen consecutive years (1950-62) the total catch from this population exceeded that level.

Signs of depletion soon became apparent. Catch rates declined, the mean lengths of both males and females in the catches decreased while the percentage of immature individuals in the catches increased. As methods were developed for determining the ages of individuals, it became clear that catches were consisting of younger individuals. The catching vessels were obliged to spend more time searching for whales and the search area around each shore station increased from year to year.

The CSIRO research group prepared a research report each year on the condition of the humpback whale stocks. These reports went to the Commonwealth Department responsible for regulating Australian whaling, and also to the Scientific Committee of the International Whaling Commission. The first warning of depletion was given in the 1953 report. This was underlined with stronger evidence in the 1954 report. Following the 1954 whaling season the Department of Commerce and Agriculture recommended to the Minister that the humpback quotas of each of the three whaling stations in Western Australia be reduced, Pt Cloates and Carnarvon from 600 to 500 whales each, and Albany from 120 to 100 whales. In the event the quota for 1955 remained at 120 whales at Albany while at the other two stations the quotas were each reduced by 100 (Table 2).

The immediate effect of the reduction in whales was that in the 1955 whaling season, gunners at west coast stations were instructed to select large whales. At one station an (unofficial) minimum length of 40 ft, was set, while another station introduced a new bonus system whereby the catching vessels were paid a bonus for every whale exceeding the previous year's average length of 39.58 ft. As a result there was a marked improvement in the size composition of catches on the west coast in 1955. However the catch per unit effort declined considerably and in fact the population size continued to fall. After the 1955 whaling season the Australian Government sold its station at Carnarvon to the North West Whaling Company which then closed its station at Pt Cloates and combined the two quotas to continue operations at Carnarvon (Table 2).

During this period the International Whaling Commission continued to maintain an overall quota for pelagic whaling in the Antarctic. In 1956 and at subsequent meetings of the Scientific Committee of the International Whaling Commission which I attended, I pressed for a separate quota for each population of humpback whales, the quota to include catches from both the breeding and feeding grounds of that stock. The Committee felt that this would be difficult to implement and that the industry would not accept such a proposition.

In the following years the research work was concentrated upon gathering the best possible documentation of changes within population structure and density. Quotas on the west coast of Australia remained at the same level from 1956 to 1959 but catches fell increasingly short of the quotas (Table 2). While it was clear that the stock was declining steadily, there was a reluctance to reduce Australian quotas while there were no parallel safeguards to that stock in the Antarctic.

From 1960 the humpback quota on the west coast was reduced progressively, but this reduction merely mirrored the decline in catch so was quite ineffective. After a disastrous season in 1963, the whaling station at Carnarvon was closed and is now the site of a fish processing works. The station at Albany had already swung to sperm whaling which has continued.

In 1963 the International Whaling Commission resolved that humpback whaling should cease in the southern hemisphere. By that time the Group IV population of humpback whales had been reduced to less than 800 individuals, more than half of which were immature.

The Group V (East Australian-New Zealand) population: 130°E-170°W

While the original size of the Group V population of humpback whales was somewhat less than that of the Group IV population, containing approximately 10,000 individuals, this population had been exposed to far less hunting during earlier years. Very few had been taken on the Antarctic feeding grounds before 1950. A shore station at Cook Strait, New Zealand had operated for a number of years but on a very modest scale. On the east coast of Australia a few humpback whales had been taken on Twofold Bay before 1930.

Thus up to 1950 this population was close to its original size. With hindsight again we now know that this population would have its maximum rate of increase at a population size of about 8000. The net rate of increase would then have been approximately 330 per year. Thus once the population was reduced to 8000 individuals, an annual catch exceeding 330 humpback whales would result in depletion of the stock. On this basis it is clear from the catches known to have been taken from this population each year (Table 3) that depletion of this stock was inevitable.

Examination of the age composition of catches taken on the east coast during the early 1950s showed that there were higher proportions of relatively old whales than were present in the population hunted on the west coast. This was in keeping with the very limited hunting of the Group V population before 1950. Because of this the Group V population did not begin to decline as early as the Group IV population.

Events on the west coast of Australia were watched closely by those whaling on the east coast. Following the reduction in quotas on the west coast after the 1954 season, stringent selection of larger whales was applied effectively on the east coast for the next five years, after which the size composition declined and more immature whales were taken.

The mean catch per hunting hour (the best indicator of density of the population) declined on the east coast from 1953 to 1956, remaining steady for the next three years and then dropped rapidly from 1959 to 1962. During this decline the searching area around each station increased each year.

Although inevitable because of the level of catching, the decline of the Group V population may have been accelerated by illegal operations of whaling fleets in that sector of Antarctic waters. It is known that 1100 humpback whales were taken there by the factory ship Olympic Challenger in 1955. Indication of later infractions came from whale marks reported as recovered from sperm and fin whales but which had been inserted into humpback whales migrating on the east coast of Australia and in New Zealand.

After the failure of the humpback whale fishery in 1962, the whaling stations at Tangalooma, Byron Bay and Norfolk Island closed. The remnant of this stock was then estimated to be close to 500 whales. As mentioned before, in 1963 the International Whaling Commission resolved that humpback whaling in the southern hemisphere should cease.

Conclusions

Management of Australian humpback whaling failed despite a sound research program and a general desire to conserve stocks. While there is little point in seeking scapegoats, there is merit in attempting to understand the reasons for this failure so that similar errors may be avoided in other fisheries. In summary the reasons appear to be as follows:
(1) As so little was known of the dynamics of these populations, Australian post-war whaling should have started with more modest catches. There should have been fewer stations and much smaller quotas for each station during the initial years. If the populations had remained stable after a number of years at the modest level of fishing, further increase in catching might then have been considered. On the other hand by commencing on a smaller scale there would have been less loss of capital if one or two stations had to be closed due to signs of stress within the populations of whales.
(2) Partly because the annual catch was far too high and partly because of a lag in identifying the key research data required, the information essential to determining realistic limits became available too late to be effective.

(3) There was also an unwillingness on the part of decision makers to accept the seriousness of the decline when warned by research workers.

(4) Even when it was clear that depletion was occurring, some argued that there was little point in Australia acting to reduce its quotas while there was not corresponding action by the International Whaling Commission to apply an overall catch limit for each population.

It should be remembered that attitudes both within Australia and internationally with regard to the regulation of whaling have been developing continuously since these difficult earlier years so that deficiencies or failures during this era cannot be used as examples of present day approaches. The specific needs of research programs are now much more clearly defined; there is much greater involvement of scientists in the management decisions, especially within the international body. Separate quotas are now set for individual stocks. Thus although the management of whale stocks may still be improved, both the approach and the results are much more effective today.

Provided that the humpback whale populations were not depleted further from those few remaining in 1963, they should not be in danger of extinction. As this species migrates so close to the shore during the breeding season, prospects for successful mating and care of young appear to be better than for species which are scattered more widely in oceanic waters during their breeding season. If allowed to recover, the Group IV and V populations of humpback whales could sustain annual catches of 350 and 300 per year respectively. However, owing to the present low levels of these populations and the relatively small net rate of increase, it may take up to 35 years for the Group IV population to reach that level and up to 50 years for the Group V population to recover.

Table 1
Catches from the Group IV population of humpback whales

Year	West coast Australia	Antarctic (70°E-130°E)	Total catch
1912	592	0	592
1913	1341	0	1341
1914	1958	0	1958
1915	1430	0	1430
1916	470[a]	0	470[a]
1925	669	0	669
1926	735	0	735
1927	996	0	996
1928	1033	0	1033
1935	0	1331	1331
1936	3072	938	4010
1937	3242	1436	4678
1938	917	866	1783
1939	0	858	858
1949	190	0	190
1950	388	779	1167
1951	1224	1112	2336
1952	1187	1127	2314
1953	1303	193	1496
1954	1320	258	1578
1955	1126	28	1154
1956	1119	832	1951
1957	1120	0	1120
1958	967	0	967
1959	700	1413	2113
1960	545	66	611
1961	580	4	584
1962	543	56	599
1963	87	0	87

(a) Includes an unknown number of sperm whales

Table 2
Quotas and catches of humpback whales from the west coast of Australia: 1949-63

Year	Point Cloates 22°35'S.113°40'E.		Carnarvon 24°53'S.113°38'E.		Albany 35°05'S.117°56'E.	
	Quota	Catch	Quota	Catch	Quota	Catch
1949	600	190				
1950	600	348	600	40		
1951	600	574	650	650		
1952	600	536	600	600	50	51
1953	603	603	600	600	100	100
1954	600	600	600	600	120	120
1955	500	500	500	500	126	126
1956			1000	1000	120	119
1957			1000	1018	120	102
1958	Transferred		1000	885	120	82
1959	to		1000	541	175	159
1960	Carnarvon		750	440	120	105
1961			475	475	105	105
1962			540	503	100	40
1963			450	68	100	19

Table 3
Humpback whales of the Group V(130°E-170°W) population: catches reported from 1949 to 1962

Year	Australian stations (quotas in parentheses)						New Zealand	Antarctic (pelagic)	Total recorded catch
	Tangalooma (27°11'S., 153°23'E.)		Byron Bay (28°37'S., 153°38'E.)		Norfolk I. (29°01'S., 167°58'E.)				
1949							141	0	141
1950							79	903	982
1951							111	162	273
1952	(600)	600					122	146	868
1953	(700)	700					109	504	1 313
1954	(600)	598	(120)	120			180	0	898
1955	(600)	600	(120)	120			112	1097[a]	1 929
1956	(600)	600	(120)	120	(150)	150	143	194	1 207
1957	(600)	600	(121)	121	(120)	120	184	0	1 025
1958	(600)	600	(120)	120	(120)	120	183	0	1 023
1959	(660)	660	(150)	150	(150)	150	318	885[b]	2 163
1960	(660)	660	(150)	150	(170)	170	361	931	2 272
1961	(660)	591	(150)	140	(170)	170	80	293	1 274
1962	(600)	68	(150)	105	(170)	4	32	0	209
Total	Australia, 8307						2155	5115	15 577

(a) Reported as being killed illegally by F.F. Olympic Challenger

(b) Redistributed after considering intermingling of populations IV and V (R.G. Chittleborough, 'Intermingling of two populations of humpback whales', Norsk Hvalfangsttid, 48 (1959), pp.510-21).

APPENDIX 13:   SUMMARY OF PRODUCTS OBTAINED FROM WHALES

Sperm oil

The oil obtained from the body tissues and head cavity of the sperm whale can be described as a liquid wax with the oil in the head cavity being particularly rich in a fraction which solidifies at room temperature. This solid fraction, called spermaceti, is usually removed by filtering.

Thin layer chromatography showed that filtered sperm oil consists of 76 per cent wax esters[1] and 23 per cent triglycerides (Hamilton et al., 1972) The esters are straight chain combinations of fatty acids and fatty alcohols which makes sperm oil unique among commercially available fatty oils.[2]   The esters present in sperm oil generally have a moderate amount of unsaturation (i.e. double bonds) and a low percentage of polyunsaturation. Chemical reaction is most likely to take place at the double bonds, which are limited in number and are placed in the ester radicals in such a way that sperm oil is very resistant to oxidative breakdown (Highgate and Job Ltd, letter to the Inquiry, 19 July 1978).

A typical distribution of acids and alcohols found in commercial sperm oil is indicated on the following page.

Sperm oil can be reacted chemically in a controlled manner due to the arrangement of double bonds.   There are two main reactions carried out by industry:
   (a)   the attachment of sulphur (or sulphur/chlorine) to the molecule at the reaction points (double bonds), producing a complex mixture of thioesters and other compounds.   The sulphur content

---

(1) Ester is the general name for the chemical compounds produced when an organic acid reacts with an alcohol.
(2) For an excellent summary of many commercially important oils see Hilditch and Williams (1964).

Acid and alcohol content of sperm oil[1]

Saturated fatty acids		%	Saturated fatty alcohols		%
Capric	(C_{10})	0.4	Myristyl	(C_{14})	3.6
Lauric	(C_{12})	2.0	Cetyl	(C_{16})	27.0
Myristic	(C_{14})	5.4	Stearyl	(C_{18})	6.0
Palmitic	(C_{16})	9.5	Eicosanol	(C_{20})	2.5
Stearic	(C_{18})	4.5	Docosanol	(C_{22})	4.3

Mono-unsaturated fatty acids			Mono-unsaturated fatty alcohols		
Dodecenoic	(C_{12})	0.7	Tetradecenol	(C_{14})	1.2
Tetradecenoic	(C_{14})	4.7	Hexadecenol	(C_{16})	7.2
Palmitoleic	(C_{16})	18.3	Oleyl	(C_{18})	48.2
Oleic	(C_{18})	37.2			
Eicosenoic	(C_{20})	12.8			
Docosenoic	(C_{22})	3.0			

Di-unsaturated fatty acid		
Linoleic	(C_{18})	1.5

Source: Highgate and Job Ltd, letter to the Inquiry, 19 July 1978

---

of the mixture can be varied but the usual range of the mixture is from 10 to 15 per cent sulphur (Furzer, 1978);

(b) the introduction of sulphate half esters (sulphated oils) or sulphonic acid groups (sulphited oils), collectively called sulpho oils (BLMRA, 1976).

Sulphurised sperm oil resulting from the first reaction has been used extensively as an antiwear additive in lubricants, particularly in demanding high temperature and pressure applications.

Sulpho oils from the second reaction are used extensively in the formulation of fat liquors for finishing leathers and cutting and drawing oils.

---

[1] For further reading on the chemical composition of sperm oil and spermaceti see Holloway (1968) and Spencer and Tallent (1973).

Sperm oil can also be converted entirely to fatty acids or fatty alcohols as required (Surmon and Ovenden, 1962). These products are used extensively in the chemical industry to produce germicides, antifoams, detergents, and numerous other products.

Spermaceti

The approximate chemical composition of pure spermaceti is outlined below.

Chemical composition of spermaceti

Free monohydric alcohols: 1-1.5%
   cetyl $C_{16}H_{33}OH$
   stearyl $C_{18}H_{37}OH$
   oleyl $C_{18}H_{35}OH$ (traces)

Free monobasic acids: 0.4%
   lauric acid $C_{12}H_{24}O_2$

Esters of monobasic acids: 98-98.5%
   Saturated: lauryl myristate (1-2%)
                cetyl palmitate (90%)
                lauryl stearate (3-4%)
                cetyl stearate   (1.1%)

   Unsaturated: unidentified (1-2%)

Source: A.H. Warth, 1947, p.83

Spermaceti has been used in the formulation of polishes, carbon paper, candles, soaps, crayons, pencil leads, release agents and food coatings.

In the cosmetic industry it is used as a wax base for hair creams, cleansing creams, antiperspiration creams, cleaning emulsions, cold creams, shampoos, facial creams, eyeshadow, lipstick, rouges, and protective skin preparations.

Whale meal and whale solubles

The solid and liquid residues from those parts of the sperm and baleen whale which have been rendered for the extraction of oil are high in protein. The solid residue, consisting largely of bone and meat pieces, is dried and hammermilled to form whale meal and the liquid residue is evaporated to produce whale solubles. Both of these products are rich in protein and are used as additives in the formulation of stock feeds.[1]

---

[1] Whale meal contains 68-75 per cent protein and whale solubles contain 92-95 per cent protein.

Sigma Chemicals Pty Ltd, a Western Australian company located in Perth, is the Australian distributor of the protein meals produced by Cheynes Beach. Forty per cent is sold to overseas markets, 20 per cent is used by farmers in Western Australia and the remainder is used by farmers in South Australia, New South Wales and Victoria and in the preparation of special vitamin-mineral-protein supplements for Australian farmers (Sigma Chemicals, 1978, and letter to the Inquiry, 19 July 1978).

Whale meal and whale solubles are two of a number of high protein stock feed additives available to Australian farmers. They are used mainly by the pig and poultry industries. 'Obviously there are substitutes for whale meal and solubles...because there are other protein sources' (Cheynes Beach, 1978, pp.33-4).

It is unlikely that any whale meal or solubles will be imported to replace the products currently being supplied by Cheynes Beach because of limited world supplies of whale meals and the competitive price of locally produced fish and blood meals and other protein supplements.

Some submissions forwarded to the Inquiry expressed concern at the high levels of mercury found in whale meal and solubles. It seems that whale meal and solubles used in pig and poultry rations at levels exceeding 1 to 4 per cent of total feed mixes, can result in mercury concentrations exceeding the 0.03 ppm National Health and Medical Research Council standard in foods for human consumption (Traverner, 1975; Plummer and Bartlett, 1975; Godfrey, 1977). Control measures directed to ensuring the whale meals are used at suitably low levels in relation to total animal dietry intake have been introduced by the Western Australian Department of Agriculture.

Whale meat

Whale meat for human consumption is taken principally from baleen whales and eaten traditionally by communities in Japan and Northern Europe in Norway, Denmark, Greenland, Iceland, Faroe Islands and within Eskimo communities bordering the Arctic circle.

Japan has the highest consumption producing some 44,000 tonnes from its own catches and importing an additional 33,500 tonnes in 1976. The Japanese eat whale meat in various forms including raw tail meat, processed sausages, canned whale ham, bacon, canned meat and vegetables and meat extracts. It is also used in processed pet foods. Whale meat accounts for 75 per cent of the Japanese whaling production (the remainder being oil and other products) and represents 4.2 per cent of Japan's consumption of animal meat (Japan Whaling Association, 1978). Whales provided about one per cent of total Japanese protein consumption in 1972 (FAO/ACMRR, 1978).

The sale of whale meat for human consumption within Australia is limited to small quantities of imported gourmet products.

Whale oil

Whale oil is the general term for the oils extracted from baleen whales. It is not therefore produced in Australia. The oil obtained from different species of baleen whales is very similar in chemical composition.

Whale oil is a true glyceridic oil consisting almost entirely of one molecule of glycerol combined with three molecules of fatty acids.

The formula may be expressed as:

$$\begin{array}{l} CH_2 - COOR' \\ | \\ CH - COOR'' \\ | \\ CH_2 - COOR''' \end{array}$$

where R', R" and R"' represent fatty acids.

The fatty acids combined with glycerol normally contain between 16 and 20 carbons and are usually different in at least two of the positions.

Hydrolysis of whale oil into glycerol and fatty acids results in the following average yields of fatty acids.

Fatty acid content of whale oil

Saturated acids		%	Unsaturated acids		%
C_{12}	Lauric	0.1	C_{14}	(-2.0 H)	2.5
C_{14}	Myristic	9.1	C_{16}	(-2.1 H)	14.0
C_{16}	Palmitic	15.3	C_{18}	(-2.5 H)	37.2
C_{18}	Stearic	2.4	C_{20}	(-7.2 H)	12.5
C_{20}	Arachidic	0.4	C_{22}	(-10.1 H)	6.4
			C_{24}	(-10.4 H)	0.1

The analytical specification for commercial whale oil is:

Saponification value	280-290
Iodine value	115-125
% free fatty acid as oleic	< 2.0
% unsaponifiable matter	< 2.0
Colour reading, Lovibond scale through 2" cell	≯ 5 Red

Source: Surmon and Ovenden, 1962, p.66

Whale oil is classified as an edible oil and is used mainly in a hydrogenated form in combination with other oils in margarine, shortening, cooking fat and soap. Glycerol and the fatty acids are used in the chemical industry. Vegetable oils and tallows and fats from domestic animals provide adequate substitutes for all uses of whale oil.

The Inquiry understands that no whale oil is imported into Australia.

Ambergris

Ambergris is formed in the stomach and intestines of sperm whales and can be found washed ashore on temperate coasts or inside sperm whales killed by whalers. In practice, however, material recovered from the animal is not held in great esteem. Tinctures and extracts of ambergris are used as fixatives in the preparation of high quality scented soaps and perfumes but ambergris is only one of a number of natural oils extracted from plant and animal origins for this purpose.

Ivory

The ivory teeth of sperm whales are used by scrimshawers for carving ornaments and jewellery.

APPENDIX 14:   JOJOBA OIL AS AN ALTERNATIVE TO SPERM OIL

Following extensive examination of the available literature and consultations with scientific experts, the Inquiry reached the conclusions presented in Chapter 7 of its report. The principal conclusion is that the oil from the jojoba shrub is a suitable substitute for sperm oil in almost all applications. However the world availability of jojoba oil is very limited and unlikely to increase significantly in the next ten years, and it would be even longer before supplies would allow prices to approach those of other lower-cost substitutes. Commercial viability could be expected while jojoba oil sells at high prices in a low volume market, but it is too early to say whether it would be profitable when competing with other substitutes in high volume applications.

These conclusions are consistent with those reached by the Commonwealth Scientific and Industrial Research Organisation in a paper on the chemical and commercial feasibility of using jojoba oil as a substitute for sperm oil, provided as part of the Commonwealth Government Agencies' and Departments' submission to the Inquiry. We have decided that this paper warrants reproduction in full and it appears below. However we have added footnotes where we believe that additional or more recent information would be helpful. We have also prepared a bibliography of those papers which we believe would be most useful to readers wishing to follow up particular aspects of the subject; this is provided at the end of the appendix.

The chemical and commercial feasibility of using jojoba oil as a substitute for sperm whale oil

The jojoba bush (<u>Simmondsia chinensis</u> (Link) Schneider) is a long-lived arid zone shrub native to the Sonoran Desert that covers parts of Arizona, California and Mexico.

The plants flower in winter and set numbers of acorn-like fruits which mature the following summer. The nuts are easily husked and the resulting seeds yield about 50 per cent by weight of a colourless oily liquid referred to as jojoba oil. The oil is chemically a liquid wax with properties unique in the plant kingdom.

It has been proposed that jojoba oil could totally replace sperm oil and spermaceti in industry. To evaluate this proposal it is necessary to examine the following questions:
- a. Given that a supply of jojoba oil was available, in what applications could it replace sperm oil and spermaceti?
- b. How likely is it that jojoba oil will become available on a large scale from either imports or Australian production?
- c. If we conclude that such imports or production are likely what would be the timetable for quantity supplies?

a. The replacement of sperm whale products by jojoba oil

<u>Chemical composition</u>: Sperm oil and jojoba oil both consist primarily of liquid wax esters (one molecule of a long chain alcohol esterified with one molecule of a long chain fatty acid). This is in contrast to both whale oil (from species other than the sperm whale) and other oils of plant origin which consist primarily of triglycerides (one molecule of glycerol esterified with three molecules of fatty acids).[1]

Unlike sperm oil, jojoba oil is remarkably pure and homogeneous in its chemical composition. The oil is simply refined by filtration through Fuller's earth which leaves it clear of all resins, tars, alkaloids or glycerides and only traces of saturated wax, steroids, tocopherols and hydrocarbons. Of the compounds present, 93 per cent are made up of C20 or C22 long carbon chains. There are virtually none of the short chain compounds which make up a high percentage of sperm oil.

Jojoba is therefore an ideal material to use unmodified as a liquid wax or in a modified form where a solid or emulsified compound is required.

---

(1) Liquid wax esters make up 97 per cent by weight of jojoba oil. Free acids and alcohols make up a further 2 per cent, with the balance being mainly steroids. In contrast with this, sperm oil contains over 20 per cent triglycerides (National Academy of Sciences, 1977).

i) Lubricants: The main use of sperm whale oil is as a component of high pressure lubricants used in gearboxes, differentials or as a cutting or drawing fluid in machine tools. The purified sperm oil is usually sulphurised but can be epoxidised, chlorinated or fluorinated before being added at the rate of 5 per cent to 25 per cent to mineral oil. Highly sulphurised sperm oil must have added mineral oil to remain liquid.

It has been verified by the USA Department of Defense (H. Gisser) that sulphurised jojoba oil shows lubricating properties equivalent to sulphurised sperm oil.[1] It has the advantage of requiring less refining in its preparation and it will remain liquid without the addition of mineral oil.

The development of this application has gone beyond the pilot study stage and lubricants containing jojoba oil are being manufactured and marketed in the USA.

ii) Solid waxes: Jojoba oil can be conveniently hydrogenated into sparkling white semi-crystalline solid wax. The pure wax is very hard with no triglyceride contamination. Softer waxes can be formulated by incorporating small amounts of other waxes or polyethylene so that the crystalline structure is weakened.

It would appear that hydrogenated jojoba oil can replace spermaceti as a wax in any industrial application including the manufacture of candles, polishes, release agents and lubricants for the bakery industry and masking and sizings in non-food products.

---

(1) National Academy of Sciences (1977) provides comparative information from various sources including Gisser, and Miwa and Rothfus. The following table is based on the work of these latter authors.

Properties of sulphurised oils from treated jojoba oil and sperm oil

Test	Jojoba	Sperm
Pour point °F	54	64
Freezing point	50	59
Flash point	491	464
Fire point	509	536
Wear test scar diameter mm	0.465	0.558
Extreme pressure, test weld, kg	240	230

Jojoba wax will also fulfill the requirements of many hard wax applications where spermaceti is not usually used.

Jojoba wax candles and hard wax products are being manufactured and sold on the USA market but the quantities involved are quite small.

iii) <u>Cosmetics and pharmaceuticals</u>: Jojoba oil is particularly well adapted to replacing sperm oil and spermaceti in cosmetics and pharmaceutical products. Because of its narrow range of chemical constituents it is relatively easy to modify the structure to produce a particular degree of hardness or a specific melting point and viscosity. This makes the material useful as a base for ointments, lipsticks and cream rouges; for aqueous creams for pharmaceutical products and antifoamants in fermentation processes.[1]

Many of these products have already been developed and are being marketed. Two brands of jojoba oil shampoo are available in the USA, and a Japanese firm is producing perfumes on a jojoba oil base. It is estimated that the requirements of the Japanese firm will reach 500 tonnes per year.

iv) <u>Minor uses</u>: Minor uses such as a low calory food oil, a leather softener and carriers for pesticides and plant hormones have been proposed.[2]

Jojoba oil reacts with sulphur chloride to form a group of compounds called factices. These compounds have potential use in varnishes, rubbers, linoleum, printing inks and adhesives.

As jojoba oil is not readily available at the moment it is possible that other substitutes for sperm oil will be developed either from mineral oils or from other plants with a shorter lead-in time. For example, work

---

(1) Partial hydrogenation of the double bond structure of jojoba oil provides further chemical flexibility in producing creams and waxes (National Academy of Sciences, 1977).

(2) A by-product of jojoba after extraction of the oil is a meal containing 26 to 32 per cent protein as well as carbohydrates and fibre. However, the potential of treating this to make it suitable as stock feed remains to be evaluated (National Academy of Sciences, 1977).

has been carried out on methods of synthesising unsaturated waxes from the glyceride esters extracted from Crambe abyssinica and Limnanthes douglasii, but results have shown that the process is uneconomic at the moment.[1]

We must conclude that if jojoba oil was readily available at a competitive price it could replace sperm whale products in almost every application.

b.   Availability of jojoba oil in Australia

i)   Imports: There is interest in jojoba in a number of countries, all of which could be potential exporters of the oil. The USA, Mexico and Israel are the most advanced in their research and in the establishment of plantations but work is also being carried out in the Sudan, Japan, Britain and Australia.

The Negev Jojoba Company of Israel is selling as much oil as they can produce in their plantations to the Japanese cosmetic manufacturers. They have plans to establish a further 400 hectares over the next four years. These plants will come into production from 1985.

There is a potential for the production of 100 tonnes per year from the wild stands of jojoba in the USA and Mexico. This represents only 0.4 per cent of the potential market calculated from the previous consumption of sperm oil in the USA.[2]

---

(1) Some discussion of this can be found in Calhoun, 1976; Anon., 1976; Nieschlag et al, 1977.

(2) 'Accurate figures are currently unavailable for the current production of jojoba oil from natural stands. The San Carlos Apache Jojoba Development Project would appear to be the largest organisation for collecting wild seed. They collected about 14 tonnes in 1977 but reported their major sale as being only 4 tonnes of oil which almost depleted their oil inventory. A large proportion of the seed is being used as nursery seed. The situation is similar in Mexico where about 10 tonnes of seed was collected last year, but most of this also became nursery seed' (Commonwealth Departments and Agencies, letter to the Inquiry, 28 July 1978).

The governments of these countries have immediate plans to establish a further 1400 hectares of plantations. However it is unlikely that production will satisfy more than 2 per cent of the potential USA market during the next ten years.

The small tonnages of oil being produced by these countries at the moment are being marketed internationally at a high price (up to $10 per kg).[1]

The American policy is to market jojoba initially as a substitute for the higher priced waxes such as beeswax and carnauba wax or for use in pharmaceuticals and cosmetics where a premium price can be obtained ($4 to $10 per kg). It is possible that when the USA market for these products is satisfied (estimated at 8750 tonnes per year) jojoba will be exported on a similar price structure rather than sold locally for low priced applications.

Thus the importation of oil from these countries would seem to be related to small tonnages sold at a premium price for specialised low volume applications. It does not seem reasonable to expect large volume imports of low priced jojoba oil within the foreseeable future.

ii) <u>Production in Australia</u>: It would appear that the only hope for a large scale supply of jojoba oil would be to develop a jojoba industry in Australia. We must therefore try to assess the potential for growth of the crop in this country.

The introduction of jojoba into Australia dates back to the mid-1930s when the late Albert Morris introduced seed from Arizona as part of a program evaluating the suitability of native and introduced species for the revegatation of denuded areas around Broken Hill. Since then experimental plantings have been established in all of the mainland states and the Northern Territory by State Departments of Agriculture or Forestry,

---

(1) A 1978 price list issued by one North American company gave prices for jojoba oil from US$17.00 to US$18.20 per kg. These prices would naturally be expected to fall substantially if jojoba oil became more readily available.

and CSIRO.[1] Generally only a few plants were established at each site, and with one exception no seed has been harvested in the field, due to a lack of male bushes in one case, and in others to ineffective pollination due to low temperatures during flowering, or immaturity of the plants. Preliminary observations on these plantings indicate that growth rates in the field can be quite low during the establishment period (10 to 20 cm per year), and that frost is likely to be a major climatic constraint on the survival or production of new plantings particularly in inland Australia. While mature bushes will recover from frosts of -5° to -9°C, seedlings are more sensitive, and the most sensitive stage of all is during flowering and pollination. One unusually heavy frost can wipe out years of work in the establishment of a new planting - a recent case was the loss of the Barona plantation of 12 hectares in California during the winter of 1975. This area was not replanted and is now considered unsuitable for the cultivation of jojoba.

Flowering occurs in response to autumn and winter rains and if the flowers are exposed to light frosts (screen temperatures of 0° to -1°C) during pollination little or no seed may be produced that year. The preferred sites for jojoba are light textured, well drained soils where long-term extreme values for screen minimums do not fall below -5°C and for stability of production the minimum should not go below zero during flowering.

In 1977 the first attempts were made to establish plantings of a hectare or more in Australia. Nursery grown seedlings were planted out that winter on at least two properties in the Victorian Wimmera and further plantings have been made for southern New South Wales and Western Australia. Because of the long lead time between sowing and full production, effective site evaluation in terms of production will not be known before 1990. Plantings on sites which are unsuitable or marginal because of soil or climatic factors will generally become apparent before then, so that by the mid-1980s we will have a better understanding of the sites on which jojoba will not grow and persist. However answers to the critical

---

(1) A Victorian company is selling jojoba seedlings for the establishment of plantations in a market campaign aimed at reaching 100 to 150 farmers during 1978 across Australia (Harms, 1978).

question about production and the stability of that production under Australian climatic conditions will not be known before the 1990s.

Studies of the natural distribution of jojoba when used to indicate suitable areas for its cultivation in other countries can be misleading. In its native habitat the criterion for success is survival. Thus for a species with a life-span well in excess of 100 years, favourable winter rains and benign temperatures are necessary only once in every 25 to 50 years to ensure adequate seed production and seedling establishment for survival of the species. In contrast the criterion for success as a domesticated crop is continuity of annual seed production. Thus its ecological range of habitats will exceed its commercial range. Also the mechanisms enabling it to survive under stress in its native habitat are generally not compatible with those necessary for sustained commercial production. Hence its domestication as a successful commercial crop will require considerable modification of the native wild types.

Controlled environment studies

Results from the phytotron in Canberra have clearly shown that jojoba will grow and flower over a wide range of temperatures ($18/13°$ to $36/31°$ - day/night temperatures) and that it is not 'summer dormant' at high temperatures when water and nutrients are freely available.

A major problem in the field is the slow establishment and the delay of at least three years before you can identify the females and adjust the ratio of male to female plants for effective pollination. Thus it is of interest that under phytotron conditions males have flowered in five months and females in nine months.

Breeding improved types of jojoba is very difficult because the plant is an obligatory out-breeder since the two sexes are separate, and the generation time is very long. New material is most likely to be selected from superior individual plants growing in plantations although we could attempt some selection from glasshouse grown material. These plants would have to be propagated vegetatively by tissue culture, cuttings or grafting. At the moment we have no plantations to select from and we will have to wait seven years before we can evaluate the plants in the plantations which will be established this year.

The use of vegetatively propagated plants would provide control of the ratio and planting arrangement of selected male and female bushes.

When sown directly into moist soil at 27°C the seed germinates within a week. The thick fleshy tap root then elongates at over 2 cm per day, and can reach a depth of 50 to 60 cm before the shoot appears above ground. Root growth following transplanting is however not so active and methods for stimulating the root growth of transplants need to be developed. While the best temperature for germination, early root and shoot development, and photosynthesis is around 27°C, photosynthesis is maintained over a very wide temperature range and does not fall below 30 per cent of the peak rate until the temperature exceeds 40°C or drops below 10°C.

It is impossible to predict the profitability of jojoba plantations in Australia at this stage as both the yields to be expected and the price to be obtained for the oil are unknown. The Israeli plantations have yielded up to 2 tonnes per hectare of clean seed at this time. This yield could increase as the plants mature. Yields of up to 4.5 kg have been measured for an individual plant. If this could be achieved for all plants in a plantation the yield could reach 8 tonnes per hectare but this is most unlikely to happen within the foreseeable future.

The price of jojoba oil is quite high at the moment, in the order of $10 kg, but this is likely to drop with increasing availability.

The cost of establishing jojoba has been estimated at around $4000 per hectare but maintenance could be as low as $250 per hectare per year. It is likely that a mechanical harvester will be developed to reduce picking costs. Since the plantation is very long lived (in excess of 100 years) the initial high cost may not be too important. In Australia many of the early plantings are being carried out on existing landholdings using the farmer's own labour and machinery so that there is little actual cash outlay.

JOJOBA BIBLIOGRAPHY

Anonymous, 1976, 'Substitutes for Sperm Oil', ACMRR/MM/SC/62, FAO Scientific Consultation on Marine Mammals, Bergen, Norway, August-September 1976

Gentry, H.S., 1958, 'The Natural History of Jojoba (Simmondsia chinensis) and its Cultural Aspects', Economic Botany, Vol.12, pp.261-95

Haase, E.F. and W.G. McGinnies (eds), 1973, Jojoba and its Uses: An International Conference, June 1972, Office of Arid Land Studies, College of Earth Sciences, University of Arizona, Tucson, Arizona

Miwa, T.K., 1971, 'Jojoba Oil Wax Esters and Derived Fatty Acids and Alcohols: Gas Chromatographic Analyses', Amer. Oil Chem. Soc. J., Vol.48, pp.259-64

Miwa, T.K., 1973, 'Chemical Aspects of Jojoba Oil, a Unique Liquid Wax from Desert Shrub, Simmondsia californica', Cosmetics and Perfumery, Vol.88, January, pp.39-41

Molaison, L.J., R.T. O'Connor and J.J. Spadaro, 1950, 'Long Chain Unsaturated Alcohols from Jojoba Oil by Sodium Reduction', Amer. Oil Chem. Soc. J., Vol.36, pp.379-82

National Academy of Sciences, 1977, Jojoba: Feasibility for Cultivation on Indian Reservations in the Sonoran Desert Region, National Academy of Sciences, Washington DC, USA

Nieschlag, H.J. et al, 1977, 'Synthetic Wax Esters and Diesters from Crambe and Limnanthes Seed Oils', Industrial and Engineering Chemistry Prod. Research and Development, Vol. 16, No.3, pp.202-7

Thomson, P.H., 1975, 'The Jojoba Nut', *West Australian Nutgrowing Society Year Book*, Vol.1, pp.6-19

Yermanos, D.M. and C.C. Duncan, 1976, 'Quantitative and Qualitative Characteristics of Jojoba Seed', *Amer. Oil. Chem. Soc. J.*, Vol.53, pp.80-82

APPENDIX 15: CORRESPONDENCE ON THE MARKET FOR SPERM OIL.

Letter from the Inquiry to Highgate and Job Ltd, 11 August 1978

From your discussions with Sir Sydney Frost and the whaling secretariat comments on prices increasing from £15 per ton in 1940 to £500 per ton in 1977 were noted.

Sir Sydney would like to make some comments in his report on the world market for sperm oil in the last three to five years. Could you assist by supplying such quarterly information as is available on sperm oil prices and quantities sold for the British and European markets since 1973? Any comments you are able to make on reasons for significant changes in price, including for example the impact of world events such as the closure of the South African operation, would be appreciated.

It would also be most useful if you could provide any indication of the likely impact on the market of the decision by the Cheynes Beach Whaling Company to cease operations in the near future.

In your discussions with Sir Sydney, mention was made of prices of various categories of substitutes up to two to four times higher than sulphurised sperm oil. I believe your company also markets a number of substitutes. It would be helpful if you could provide indicative prices for recent years highlighting the differentials between filtered sperm oil, sulphurised sperm oil and various types of substitutes used in leather and lubricant applications.

Your assistance in this matter would be greatly appreciated.

Letter to the Inquiry from Highgate and Job Ltd, 22 August 1978

From the text of your message it would appear that you are endeavouring to create a summary of the movements in Sperm Oil over the last ten years and this I shall try to give you.

It may not be complete but I trust it will enable you to form an estimate of the situation over the period under review.

For this purpose, I enclose two Schedules -
(1) A Schedule of the variation in price of Crude Sperm Oil over the years 1970-1978, divided into quarterly periods.
(2) A Schedule of Crude Sperm Oil Imports to the United Kingdom for the years 1974-77. (It is appreciated that in your Telex you ask for Europe. This we are unable to do and hope that these figures for the United Kingdom will be of some assistance to you.)

Schedule I is obtained from an examination of price structure retained in this Office, while the prices and tonnages shown in Schedule II are obtained from Customs Returns. In consequence a direct comparison of price structure cannot be inferred one to the other.

In commencing with the year 1970 we are really beginning in a new cycle of price structure. During the preceding ten years the price variation was from £45 per tonne up to £120 and back to £90. This covered a normal sequence with the £120 per tonne price being, at that time, considered a very excessive peak, at which stage tonnage began to be lost by the users showing resistance.

However, with the introduction of the seventies and the inflationary period that followed, a new sphere of prices naturally ensued.

Between 1970 and 1975 the price modulated between £90 per tonne and £175 per tonne, which was a fairly constant figure when placed against supply and demand and this structure was favourably accepted by the end users of Sperm Oil. It was acceptable at that time as it showed an advantage over the synthetic prices ruling on that day.

Between 1976 and 1977 an exceptional upsurge occurred in the price of Crude Sperm Oil ranging from the £170 operating in 1975 to £320 in 1976 and £470 in 1977. This was an unfortunate sharp increase and it was due to several factors -
(a) The I.W.C. in June, 1976 severely reduced the quotas for the southern hemisphere resulting in a panic position being created.
(b) Due to this reduced tonnage South Africa decided to cease production.
(c) At this time the cost of mineral oil rose sharply, thus increasing the costs of production to the whalers.

All these factors tended towards this sharp increase which in the end proved to be the undoing of the Sperm Oil Market as a very strong consumer resistance was set up.

In 1977 and 1978 a reactionary movement took place due to end users' resistance and the price has now fallen to about £250 per tonne.

At this revised level consumers are very interested in the use of this Oil but due to the uncertainty of the situation over the preceding years, a favourable market no longer exists as the raw material has lost interest with those who depended upon it most.

Due to this uncertainty a trend has been created to investigate the use of synthetics even though it is recognised that the synthetic material cannot produce a finished product equal to that from Sperm Oil.

It is very indefinite as to the effect of the withdrawal from production of South Africa, due to the other conditions which were prevalent at that time. In consequence no comparison can be made here with the effect which the Australian withdrawal this year will have on the crude market.

As is known to you, the waters of area 5 of the southern hemisphere are still open to the fishing of Sperm Whales by any other whaling country, most likely Japan and Russia. In consequence, the catch from these waters may not be reduced by the withdrawal of Australia from the Sperm Oil enterprise. In so saying, however, it is recognised that this could greatly be affected by any legislation passed by the Australian Government and accepted by International Marine Authorities, by which foreign countries are no able to fish within a prefixed mileage limit of the Australian coastline.

These points are of great uncertainty and may best be answered by the legislators of Australia itself.

It is anticipated that Sperm Oil will still be available from Japan, Iceland, Peru and perhaps Chile so that the **international** market will still be served from these sources.

With the falling off of supply it is considered that there will be a slight diminution of demand. As has been prevalent over the last five years, the supplies of Sperm Oil available are now finding their way into the hands of those who require it most. This is the ideal situation which should be encouraged.

In this way, it is our opinion that the demand for Sperm Oil will continue in a much more selective basis than heretofore. It is, therefore, difficult to assess what effect the withdrawal from Australian waters by Cheynes Beach will have. However, it would be advantageous for levelling out the supply position if this activity could be maintained.

With regard to the last paragraph of your message concerning substitutes, these substitutes are derived from various sources, some being from the use of natural oils such as animal oils and others from prescribed special chemicals. Due to the variations in their creation so also does the price vary and it may be assessed that the price of substitutes would range from £400 to £800 per tonne.

The synthetic materials or substitute materials at the lower end of the scale are greatly in favour but the products produced from them are greatly inferior to those when Sperm Oil is used, and therefore they can only be used in limited lines of products, whereas those at the upper end of the scale, namely, £800 per tonne, which give a product as equal to that of Sperm Oil as possible, are considered too high for the purpose and therefore we have a comparison of price against quality.

One further Oil which is being considered internationally is the Jojoba Oil but as this is not yet obtainable in commercial quantities, it is considered uneconomic to be recognised as a positive raw material. However, if this material could be produced at this moment in quantity as a replacement for Sperm Oil, it is considered that it would cost about £3,250 per tonne.

In a previous paragraph the reference to the lower range products may be evidenced by lubricating materials for cutting oils, whereas the higher range products are most likely to be found in the Leather Industry.

Yours faithfully,
HIGHGATE & JOB LTD

P.S. The price of £15 per tonne in 1940 now relates to a forgotten age.

SCHEDULE I

Crude sperm oil prices

	Quarter			
	1	2	3	4
		£		
1970	95	125	125	140
1971	155	150	140	115
1972	115	105	115	125
1973	120	140	155	150
1974	180	205	185	185
1975	180	175	175	150
1976	230	320^a	340	370
1977	390	420	470	400^b
1978	275	250		

(a) At this point the quota was cut and the excessive increase arose.
(b) The stage at which end users' resistance became most evident.

SCHEDULE II

Schedule of crude sperm oil imports to the United Kingdom for the years 1974-1977

YEAR	QUARTER 1		QUARTER 2		QUARTER 3		QUARTER 4	
	Tonnes	Price per tonne	Tonnes	Price per tonne	Tonnes	Price per tonne	Tonnes	Price per tonne
		£		£		£		£
1974	2048	108	3103	145	1005	160	2594	180
1975	1878	205	1309	209	3402	182	1208	154
1976	2076	173	3229	166	2010	246	1213	486
1977	2022	415	903	565	2227	503	2259	407

Letter to the Inquiry from Cheynes Beach, 23 November 1978

We have had the opportunity of reading the letter from Highgate and Job Limited of 22 August 1978. We will for the purpose of our discussion of the market accept the crude sperm oil prices as set out in Schedule I. While our own experience shows some slight variation on these prices we believe for discussion purposes that they are satisfactory.

General market position in crude sperm whale oil: There has been over the many years of this Company's operation continual movement of the price of crude sperm whale oil. This movement is sufficient to change the economics of operating a whaling industry from profitable to unprofitable This is an accepted fact of life for those operating in this market and the following commentary on the market should be read on the understanding that whaling operators look to a longer term view when dealing with depressed pric conditions for their products and this attitude has enabled the industry to continue on a profitable basis.

There have been two significant factors which have affected the market for crude sperm oil and they are:
    (a) The reduction in supply of crude sperm oil by
        (i) Russia using all its production internally and
        (ii) the more recent reduction in quotas as determined by the International Whaling Commission.
    (b) World-wide inflation.
Both these factors have caused crude oil prices to increase.

As would be expected some of the end users of processed crude sperm oil have shown a resistance to the higher prices. This Company regards that resistance as being a natural and necessary market reaction for balancing supply and demand for the product.

The Company saw as a more significant threat to the continuation of a stable whaling industry the possibility that supply might be reduced to such a low level that processors would not be able to obtain sufficient of the product to supply their end users on an economic basis. Thus the potential reduction in supplies of crude sperm oil and the uncertainties of continuity of supply which were developing in the market were seen as the most significant threat.

The Company, to overcome this foreseeable difficulty, was in the process of negotiating direct with a major processor to arrange a joint venturing to ensure that the supplies from our Company (being a major supplier to the free market) could be used to stabilise the industry and overcome much of the uncertainty which was developing in the market in relation to the continuity of supply.

This proposal was considered viable by the Company and by the party with whom we were negotiating and it was considered to be profitable to all concerned. As part of the proposal the processor of the crude oil was to spend approximately A$300,000 to provide modern processing and refining equipment.

The uncertainty in relation to continuity of supply was reinforced by the American blockade of the Company's oil shipment which occurred in 1977. While the difficulties that arose were due entirely to the American Government's refusal to allow transit of the oil in any ship passing through one of its ports, much credit for the delay was taken by the conservation movement. In fact the conservation movement played no part in the delays occasioned and the matter was finally resolved on a normal commercial basis.

The result, due to the wide publicity given to the issue, was seen as another unsettling factor within the market and further accentuated the resistence of the buyers.

Crude sperm oil prices: We will limit our discussion on crude sperm oil prices to those occurring from 1973 to 1978. As a result of the matters referred to above the supply of crude sperm oil on a world wide free market basis had reduced and the price in consequence was beginning to rise from 1973 to 1974.

In 1975 a major trader in crude sperm oil in Europe went bankrupt while holding approximately 8 - 10,000 tons of crude sperm oil. The bankruptcy was unrelated to the crude sperm oil being rather as a result of trading in other commodities. Nevertheless the result was that approximately 8 - 10,000 tons of crude sperm oil could have been released on to the market in an uncontrolled fashion. The natural effect of this position was to depress crude sperm oil prices as indicated by the 1975 figure in Schedule I of the letter of Highgate and Job Limited of £150.

In the Company's opinion this depressed market position was against the price cycle which producers have come to expect in the crude sperm oil market and as a result when this problem was resolved prices once again began to increase. The resulting increases which appeared to be substantial were in fact a taking up of the increase which would otherwise have occurred in 1975 plus the natural market movement. The Company recognises that the fact that prices rose very sharply over a short period did have an unsettling effect on the end users of refined crude sperm oil.

Thus to put the market in a perspective at the beginning of 1978 it is the Company's view that the market was in an unsettled state as the result of a significant price rise over a short period, the difficulties created by the American Government's embargo on the transit of the crude sperm oil shipment in 1977 and a topping off of the upward movement of price increases for crude sperm oil.

It was widely anticipated in February of 1978 that, as had happened many times before, the demand for crude sperm oil would once again show some improvement by mid-1978 and provided the uncertainty of supply which was prevalent in the market could be contained the market would continue.

In early 1978 the Australian Government announced its Inquiry into Whales and Whaling with one of the terms of reference being whether Australian whaling should continue or cease.

The importance of this Inquiry became more apparent to the European market as the Inquiry made itself known to market operators and requested information of them. This Company assisted the Inquiry in making contact with many of those operating in this market.

In the Company's view it was an unforeseen effect of the Inquiry, probably brought about by the unique set of circumstances previously mentioned which were operating at the time which caused the market to react adversely to the possibility of Australian whaling ceasing. The resulting further uncertainty within the market was the straw that broke the camel's back.

It was considered necessary for the Company to announce the cessation of its whaling operations to enable a sale to be negotiated for the oil which was then on hand and if possible for the production of the remainder of the year. As anticipated by the Company the announced closure sparked sufficient interest in the oil we were currently holding to enable a sale to be negotiated but at a greatly reduced price of US $510.

Future oil supplies: The whales which were taken by the Company were the result of a quota for Division 5. On the basis that a quota remains for Division 5 the quota will go back into the international pool and be divided up between whaling nations. As Division 5 extends over a huge area of ocean not necessarily controlled by Australia it is reasonable to expect that one of the major whaling nations will take up the quota.

It is the Company's understanding that the Japanese have in previous years traded off sperm whale quotas for quotas in other edible whales. If this should happen to the increased quota that the Japanese are likely to obtain from the cessation of Australian whaling then the Russians will take up the sperm whale quota and the oil produced will be used internally within that country and not form any part of the free market supply.

Cessation of whaling by South Africa: South African Whaling Company ownership, just prior to its cessation of whaling, had undergone many changes. It is our understanding that with each change of ownership the Company's resources were directed to new activities with the result that the reduction in resources eventually applied to the whaling activities made its continuation impractical.

No doubt variations in the market had some effect. However as previously mentioned whaling operators accepted downturns as a normal part of the market movement and this would not necessarily have been sufficient to cause a cessation of whaling.

Yours faithfully,
CHEYNES BEACH WHALING COMPANY (1963) PTY LTD

APPENDIX 16: WHALE BRAIN AND ITS INTELLIGENCE POTENTIAL

(A submission to the Inquiry by Myron S. Jacobs)

General statement on brain and intelligence

In all animals with backbones, the brain is the chief organ of the central nervous system (brain + spinal cord). The pre-eminent position of the brain has apparently resulted from processes that have acted on the nervous system of animals throughout evolution. With the appearance and ascendance of mammals in the animal Kingdom, the brain has become progressively larger. This has been accomplished primarily by the addition of more neural tissue to the cerebral hemispheres (neocortex) and cerebellum (neocerebellum). It is the great surface mass of neocortex in the human brain, as compared to other land mammals, that is generally accepted as the basis for man's ability to ideate, symbolise, create language and develop implements to manipulate his environment.

The whale is endowed with a neocortex that is even more abundant than that of man and for that matter any other mammal. Despite this common neural basis for intelligence, the extreme dissimilarities of environment and the sound communication barrier have, so far, stood in the way of general acceptance of the view that whale and human brains share the same functional capabilities.

Status of whale brains with respect to mammalian evolution

Approximately 25 million years have elapsed since the land ancestors of modern whales made the successful and progressive re-entry to the totally aquatic environment. Fossil records of the early whale forms (Archeoceti) are fragmentary and limited. The evidence suggests, however, that ancient whales were small compared to today's great whales and probably had small brains comparable to those of the descendants of their immediate insectivore-like terrestrial precursors.

It should be emphasised, when making comparisons between the brains of whales and humans, that environmental factors have been acting on whales for a much longer period of time during their evolution than on hominid forms. Probably of even greater importance is the marked differences in the type, and changeability of the environments in which whales and man have developed during their separate evolutions. The terrestrial environment was subjected to cataclysmic upheavals associated with periods of glaciation and reorganisation of land masses. These severe environmental fluctuations very probably played a major role in determining the selective pressures that moulded the form of the evolving human body and brain and in the development of language as a means of human communication. During these periods of upheaval, the terrestrial environment was subject to great alterations in the availability and type of food, in the accessibility to potable water, in extremes of ambient temperature and in the stability of the earth's crust. On the other hand, owing to the physical properties of water, the oceans were relatively less affected. It is reasonable to assume, therefore, that whales have been less subjected to major environmental variables and that the selective pressures influencing their body and brain development in the aquatic environment were more constant once beyond the critical early segment of their evolution when the transition from land to water was accomplished.

It must be assumed that the long period of isolation from terrestrial lines of mammalian descent totally immersed in an environmental medium so different from that which surrounds land and aerial mammals was responsible for the changes that have occurred in the body form of whales. Some of these changes in body form have been of greater importance than others in influencing the gross shape of the brain and also certain of its functions. Thus, in whales, modifications in the bones of the facial skeleton have occurred in connection with reorienting the nasal passageways to open onto the body's surface in the forehead region as the blowholes rather than in the midregion of the face as the external nares. This reorganisation of the facial skeleton has been responsible for reducing (in the baleen whales) or eliminating completely (in the toothed whales) the bony orifices through which olfactory nerves pass from olfactory endings in the nose to reach the brain. This evolutionary alteration resulted in the partial or total loss of the olfactory function. (Experimentally, the same result can be demonstrated following surgical destruction of the olfactory nerves in osmatic animals.) Additional modifications of the bones at the base of the skull also occurred during evolution in association with shortening of the neck region. Collectively, these modifications in the bones of the head have been responsible for creating in whales a cranial cavity that is considerably more foreshortened, as well as higher and wider, than the cranial cavities present in land mammals. The shape of the brain that develops within a cranial cavity is greatly influenced by the shape of its bony covering. Thus, the cetacean brain has assumed an expanded globular shape; that is, one that is higher and wider than it is long. The latter configuration is more typical of the brains of terrestrial mammals, including man. In connection with the altered shape of the brain in modern whales, the brain has become markedly flexed and also rotated forward and downward to an 'on-end' position within the cranial cavity. This latter rotation appears closely related to the appearance of an additional flexure of the brain stem interposed between the expanded rostral end of the brain (telencephalon consisting primarily of the cerebral hemispheres) and the spinal cord.

Of greater significance than the remodelling of the head bones, in considering the functional status of the whale brain, is the effect that the aquatic environment seems to have had during evolution on emphasising development of the head region. The massiveness of the head of the sperm whale is the best example of this trend. In contrast to richer and more variable sensory inputs provided land mammals by gravity, contact with an unyielding environmental substrate, diurnal and nocturnal light cycles and seasonal, climatic and temperature fluctuations, whales have evolved in a relatively stable, gravity neutral, opaque, acoustically rich and yielding environment. If these latter ambient conditions were all that nature required to produce large-headed animals in the oceans, then one might expect to see this feature generally throughout the phylogenetically more ancient teleost and elasmobranch fish. However, even the largest of the grouper fish and sharks have relatively small heads.

The large size of the head in whales is paralleled by a massive enlargement of the brain. Only a part of this brain enlargement appears to be due to its greatly increased sensory inputs through the cranial nerves, especially the cochleo-vestibular and trigeminal, and also its increased motor outflow to head structures, especially through the facial nerve. The bulk of the increased brain size is the result of a bilateral expansion of the two cerebral hemispheres, most of which are covered by neocortex. It is the great expansion of the hemispheres that sets mammals apart from other animals behaviourally and enables them to respond with more adaptability and flexibility to new situations. In man, the increased volume of neocortex, present in functional linkage with an adequate 'vocal' tract, has made possible the development of language and linguistic communication.

While it is uncertain just how far one can validly extend functional extrapolations from one mammalian brain to another, the impressive volume of neocortex in the human brain, especially neocortex of the associational type is duplicated in the whale. Neural similarities aside, the air passageways in whales are organised differently than in man owing to the absence of vocal folds in many whales and the imperative nature of preventing water from being inspired while swimming beneath the surface. Nevertheless, movement of trapped air between air sacs does occur in cetaceans and is responsible for a wide variety of sounds emitted by these animals. As a group, toothed whales are noisier in this respect than baleen whales. Behavioural studies with dolphins indicate that their sound emissions are used for social interactions, as well as for echolocation.

It is not known with any certainty whether, in the former context, the sound emissions represent a cetacean counterpart of language. Certain isolated bits of anatomical and behavioural evidence suggest the possibility that vocal exchanges between dolphins may, in fact, constitute a special type of 'language' that has developed in isolation in the oceans. In this connection, it has been only within recent years that behavioural studies with subhuman primates have demonstrated that chimpanzees are capable of making symbolic cortical associations, an activity that underlies thought processes and language. Moreover, the brain size of most whales and dolphins considerably exceeds the minimum 900 gm mammalian brain weight considered to be critical for supporting a language function.

Statements on brain functions

No comparative neurologist would disagree with the fundamental fact that in <u>all</u> mammals, the brain functions (a) to receive and process sensory information from the external environment, from the surface of the body and from the internal environment within the body, (b) to initiate, direct and influence somatic and visceral motor activities, and (c) to modulate or control muscular activity so that body movements are smooth and well timed.

In its role as a sensory analyser, the generalised mammalian brain receives its information directly through certain of twelve pairs of nerves that enter the cranial cavity (cranial nerves), and indirectly through paired nerves that enter the spinal cord (spinal nerves). Some of the cranial nerves are responsible for transmitting to the brain information from specialised sensory endings in the head that respond to specific forms of energy in the outside world. Such are the <u>optic nerves</u> which mediate visual and non-visual photic inputs and the <u>cochleo-vestibular</u> (auditory) nerves mediating auditory and vestibular inputs. (Vestibular functions relate to the position of the head in space and to angular and linear movements of the head.) In addition, the <u>olfactory nerves</u> (absent in adult toothed whales and reduced in baleen whales) convey to the brain information regarding environmental odours, and <u>gustatory nerve fibres</u> in up to three of the cranial nerves (<u>facial</u>, <u>glossopharyngeal</u> and <u>vagus nerves</u>) transmit information to the brain regarding taste.

Other nerves convey information from sensory endings that respond to stimuli acting <u>directly</u> on the somatic mass of the head and body. Such are the <u>trigeminal nerve</u> of the head and all of the spinal nerves that transmit information indirectly to the brain from sensory endings in the body that respond to tactile, vibratory, pressure, thermal and painful stimuli. Certain of these endings and associated nerve fibres also mediate what may be referred to as 'position sense', that is awareness of the relative position of dependent parts of the body to environmental space.

The third activity in which the brain serves as a sensory analyser relates to those inputs arising essentially from the visceral environment within the body. Such information is related to states of distention of visceral organs, levels of glandular secretion and degrees of contractility of the tubular visceral organs. This information is transmitted to the brain directly through the <u>facial</u>, <u>glossopharyngeal</u> and, particularly, the <u>vagus nerves</u>, but also indirectly through some of the spinal nerves.

In its second major function, that of initiating and controlling somatic and visceral motor activity, the brain serves to project its influence either directly or indirectly upon neurons (nerve cells) within the central nervous system that activate peripheral motor structures such as muscles and glands. These peripheral motor activities are mediated by fibres (axons) of the aforementioned central neurons that leave the central nervous system and pass into the periphery as motor fibres of all spinal nerves and of nine of the twelve pairs of cranial nerves. Rather limited regions of the neocortex are directly responsible for initiating voluntary movements.

In contrast to the rather limited amount of neocortex that can initiate movements, all regions of the neocortex, as well as the cerebellum and certain other brain structures, play an important role in the third major function of the brain, namely, fine control of the body's movements. This is effected by a number of neuronal loops or circuits that serve to regulate the activity of neurons responsible for initiating motor activity. By regulating both the rate at which these neurons are discharged and also the number of neurons that are activated at any moment, individual muscle contractions and group muscle actions are finely controlled. By this type of exquisite control, the dependent parts of the body are moved in a precise sequence and through well timed trajectories.

The highest levels of brain function relate to its ability to ideate, symbolise and create language. The neural mechanisms underlying these activities are still unclear, but they certainly involve the association cortex as well as other cortical regions, perseveration of neuronal activity, preferred cortical pathways and neurotransmitters. These activities, until recently, were generally regarded as being possible only in the human brain. It is obvious that such a point of view would logically preclude the possibility of humanoid intelligence or language existing in any other species.

The brain functions considered above are responsible for the characteristic behavioural attributes that, in man, have enabled the species to dominate its environment and, by encroachment into the environmental spheres of other species, to influence and modify the stability of numerous ecological systems. Whales, too, appear to have come to dominate their special world. This mastery, occurring in an endless fluid environment that does not lend itself readily to manipulation by hands or tools, must be attributed to the great size of their brain and body.

Statement on neuronal basis of brain function

In all of the functions just described, the brain serves as a decoder of the bioelectric inputs (impulses) that it receives. Among the untold billions of neurons that collectively form the nervous system, any single neuron is capable of responding to a suitable stimulus in only one of two ways. It is either excited so that the impulse that is generated is transmitted to the ends of that neuron, or it is inhibited from giving rise to an impulse for a fleeting instant. In a sense, neurons function like the basic units of a binary computer that can respond in only one of two ways.

Other than in the cerebral cortex, the neurons of the central nervous system tend to be organised in series, with varying numbers of junctional gaps (synapses) present between successive neurons in any particular neuronal pathway. Although the responsiveness to stimuli and the generation of impulses in individual neurons is stereotyped, the speed at which neurons transmit their impulses, and therefore the rate at which they can be activated, varies considerably. Moreover, the degree of branching (collateralisation) of individual neurons varies from one to another of the so-called functional systems of neuronal pathways. It seems, therefore, that the functional significance which the brain attaches to the bioelectric impulses it receives from any specific neuronal pathway is based on the total pathway, the area of the brain receiving the information, and also on the time sequence (temporal pattern) and three-dimensional pattern of neurons activated and inhibited in the cerebral cortex (spatial pattern).

The cerebral cortex has a different cellular organisation. In general, its neurons are organised both in series and parallel. Cortical neurons are arranged in series radially across the thickness of the cortex where they form 'pencil-like' columns of cells extending from the superficial surface to the deepest zone of the cortex. In addition, these serially arranged columns of neurons are interconnected in parallel by collateral and terminal fibres which structurally and functionally link the cells of the cerebral cortex into a three-dimensional system. The parallel arrangement links not only nearby radially oriented series of neurons, but also radially oriented neurons of different lobes of the brain. The linking together of different regions of the brain having to do with different functional activities is the anatomical basis of the so-called association areas of the cortex which are so highly developed in both whale and man.

Summary statements

Three facts are of cardinal importance when comparing the functional capacity of the whale brain to that of man. First this magnificent line of mammals has been evolving for a considerably longer period of time than man. Second, adaptations and specialisations occurring during prolonged isolation in the aquatic environment have resulted in a body form that is entirely different from man's, one that is reflected in adjustments in the organisation of the central nervous system. Third, despite the prolonged separation from terrestrial lines of mammalian evolution, whales have brains that are quite comparable in size and complexity to that of man, a state not achieved by any other aquatic mammal or, in terms of size, by any primate. In view of these facts, whales as a group should be thought of as having reached a level of morphological and functional brain development in the aquatic environment comparable to man's in his environment.

APPENDIX 17: AUSTRALIAN CONSERVATION FOUNDATION
OPINION POLL ON WHALING

The following is a summary of results of the opinion poll.   %

Q1. If whales are an economic source of raw materials, we should continue to kill them within the limits scientists consider would not endanger the species 1. 34.9

OR

Even if it costs more to obtain products which currently come from whales, it would be better to pay the extra cost rather than kill them 2. 58.0

Don't know/not established 3. 7.1

Q2. To catch whales may lead to their becoming extinct even if we limit the number which may be caught in any one year 1. 37.0

OR

Scientists can be relied upon to tell us how many whales may be caught each year without any risk of their becoming extinct 2. 54.3

Don't know/not established 3. 8.6

Q3. The people in Australia who earn a living from whaling have a right to continue to earn their living in this way 1. 27.7

OR

People in the whaling industry should be helped to retrain for other work at the community's expense 2. 63.1

Don't know/not established 3. 9.2

Q4. It would be stupid for Australia to ban whaling if other countries which are whaling were to continue 1. 29.7

OR

	%
Australia should follow the example of other countries which have banned whaling, even if some countries decide to continue .....2.	61.4
Don't know/not established .....3.	8.9

Taking all things into consideration, do you believe:

Q5. Whales should not be killed at all, even if it could be shown that whaling does not threaten the existence of the species 1.   41.6

OR

Whaling should be continued on a controlled basis 2.   51.8

Don't know/not established   6.6

BIBLIOGRAPHY

ACMRR/MM/SC/5 Add.1, 1976, Addendum to Draft Report of Ad Hoc Group IV Ecological and General Problems, Scientific Consultation on Marine Mammals, FAO (see FAO/ACMRR, 1978)

Allen, K.R., 1973, 'The Computerised Sperm Whale Model' in <u>International Whaling Commission Twenty-third Report</u>, London, UK, pp. 70-4

_____, and J.L. Bannister, 1977, 'Whaling: Then and Now', Environs, 1:5, pp. 8-9

_____, and D.G.Chapman, 1977, 'Whales', in J.A.Gulland (ed.), <u>Fish Population Dynamics</u>, John Wiley & Sons, London, pp.335-58

_____, and G.P.Kirkwood, 1977(a), 'Further Development of Sperm Whale Population Models', in <u>International Whaling Commission Twenty-seventh Report</u>, Cambridge, UK, pp.106-12

_____, and G.P.Kirkwood, 1977(b), 'A Computer Program to Calculate Sustainable Yields and Fishing Mortality for Sperm Whale Populations (SPVAP) of Variable Size', in <u>International Whaling Commission Twenty-seventh Report</u>, Cambridge, UK, p.263

_____, and G.P.Kirkwood, 1977(c), 'Program to Calculate Time Series of Sperm Whale Population Components for Given Catches (SPDYN)', in <u>International Whaling Commission Twenty-seventh Report</u>, Cambridge, UK, p. 264

_____, and G.P.Kirkwood, 1977(d), 'Were 1946 Sperm Whale Stocks at the Unexploited Level?', Report No. 87, CSIRO Division of Fisheries and Oceanography, Sydney

Anonymous, 1976, 'Substitutes for Sperm Oil', ACMRR/MM/SC/62, Scientific Consultation on Marine Mammals, FAO, Rome, (see FAO/ACMRR, 1978)

Ash, C.E., 1964, 'British Whaling: Final Years', Chemistry and Industry, Vol. 38, Pt 2, pp. 1596-1601

Bannister, J.L., 1974, 'Whale Populations and Current Research off Western Australia', in W.E. Schevill (ed.), The Whale Problem: A Status Report, Harvard University Press, Cambridge, USA, pp.239-54

_____, 1978, 'Whale stocks off Western Australia - a Review', Submission to the Inquiry into Whales and Whaling

Berzin, A.A., 1972, The Sperm Whale, Israel Program for Scientific Translations Ltd, Jerusalem

Best, P.B., 1970, 'Exploitation and Recovery of Right Whales off the Cape Province', Investl. Rep. Div. Sea Fish. S. Afr., Vol.80, pp.1-20

_____, 1974, 'The Biology of the Sperm Whale as it relates to Stock Management', in W.E.Schevill (ed.), The Whale Problem - A Status Report, Harvard University Press, Cambridge, USA., pp.257-93

_____, 1975, 'Death-times for Whales Killed by Explosive Harpoons', in International Whaling Commission Twenty-fifth Report, London, UK, pp.208

_____, 1976(a), 'A Review of World Sperm Whale Stocks', ACMRR/MM/SC/8 Rev.1, Scientific Consultation on Marine Mammals, FAO, Rome, (see FAO/ACMRR, 1978)

_____, 1976(b), 'Status of Whale Stocks off South Africa, 1974', in International Whaling Commission Twenty-sixth Report, London, UK, pp.264-85

_____, 1978, 'Social Organization in Sperm Whales, Physeter Macrocephalus' in H.E.Winn and B.L.Olla (eds), Behaviour of Marine Animals, Vol. 3, The Natural History of Cetaceans, Plenum Press, New York, pp.231-93

BLMRA, 1976, Utilization of Sperm Oil by the Leather Industry, a report by the British Leather Manufacturers' Research Association for the British Department of Industry

Bryden, M.M., 1978, 'The Status of Humpback Whales of Area V, 1977' (mimeo.), interim report to the Australian National Parks and Wildlife Service

Bunnell, S., 1974, 'The Evolution of Cetacean Intelligence', in J.McIntyre (ed.), Mind in the Waters, Charles Scribner's Sons, New York, pp.52-66

_____, 1978, 'Whale Brains and Human Ethics', Submission to the Inquiry into Whales and Whaling

Bureau of International Whaling Statistics, International Whaling Statistics, Vol. I (1930) to Vol. LXXX (1977), Sandefjord, Norway

Caldwell, D.K., 1978, Letter to the International Whaling Commission, in IWC/SC/30/Rep.5

Caldwell, M.C. and D.K. Caldwell, 1966, 'Epimeletic (Care-giving) Behaviour in Cetacea', in K.S.Norris (ed.), Whales, Dolphins and Porpoises, University of California Press, Berkeley and Los Angeles, pp.754-89

Calhoun, W., 1976, 'Limnanthes alba (Meadow Foam) Oil as a Potential Substitute for Sperm Whale Oil', ACMRR/MM/SC/56, Scientific Consultation on Marine Mammals, FAO, Rome, (see FAO/ACMRR, 1978)

Cheynes Beach Whaling Co. (1963) Pty Ltd, 1978, Submission to the Inquiry into Whales and Whaling

Chittleborough, R.G., 1956, 'Southern Right Whale in Australian Waters', J. Mammal., Vol. 37, pp.456-7

_____, 1959 'Australian marking of Humpback Whales', Norsk Hvalfangst-tid Vol. 2, pp.47-55

_____, 1978, 'Australian Humpback Whaling', Submission to the Inquiry into Whales and Whaling

Clark, C.W., 1975, 'The Economics of International Whaling', in Monitor, 1978, Submission to the Inquiry into Whales and Whaling

Clarke, M.R., 1970, 'Function of the Spermaceti Organ of the Sperm Whale', Nature, Vol. 228, pp.873-4

_____, 1978(a), 'Structure and Proportions of the Spermaceti Organ in the Sperm Whale', J. Mar. Biol. Ass., UK, Vol. 58, pp.1-17

_____, 1978(b), 'Physical Properties of Spermaceti Oil in the Sperm Whale' J. Mar. Biol. Ass., UK, Vol. 58, pp.19-26

_____, 1978(c), 'Buoyancy Control as a Function of the Spermaceti Organ in the Sperm Whale', J. Mar. Biol. Ass., UK, Vol. 58, pp.27-71

Clarke, R., 1952, 'Electric Whaling', Nature, Vol. 169, pp.859-60

Colwell, M., 1969, Whaling Around Australia, Rigby Ltd, Kent Town, South Australia

Commonwealth Government, 1978, Commonwealth Government Agencies' and Departments' Submission to the Inquiry into Whales and Whaling

Connecticut Cetacean Society, 1978, Submission to the Inquiry into Whales and Whaling

Dakin, W.J., 1977, Whalemen Adventurers in Southern Waters, A & R non-fiction classics edition, Angus and Robertson, Sydney

Darwin, C., 1883, The Descent of Man, 2nd ed., John Murray, London

Dawbin, W.H., 1959, 'New Zealand and South Pacific Whale Marking and Recoveries to the End of 1958', Norsk Hvalfangst-tid, Vol. 5, pp.213-38

_____, 1966, 'The Seasonal Migratory Cycle of Humpback Whales', in K.S.Norris (ed.), Whales, Dolphins and Porpoises, University of California Press, Berkeley and Los Angeles, USA, pp.145-70

FAO/ACMRR, 1978, (Draft) Proceedings of the Scientific Consultation on the Conservation and Management of Marine Mammals and their Environment, Bergen 1976, to be published by the Food and Agriculture Organisation of the United Nations

FAS, 1977, Animal Rights, Federation of American Scientists, Public Interest Report (Special Issue)

Furzer, I.A., 1978, Submission to the Inquiry into Whales and Whaling

Gambell, R., 1976, 'World Whale Stocks', Mammal Review, Vol. 6, No. 1, pp.41-53

Gaskin, D.E., 1976, 'Evolution, Zoogeography and Ecology of Cetacea', Oceanogr. Mar. Biol. Ann. Rev., Vol. 14, pp.247-346

Godfrey, N.W., 1977, 'Replacement of Meat and Bone Meal with Whale Meal or Whale Solubles in Pig Diets and the Accumulation of Mercury in the Carcass', Australian Journal of Experimental Agriculture and Animal Husbandry, Vol. 17, pp.403-11

Godfrey-Smith, W., 1978, 'The Ethics of Whaling', Submission to the Inquiry into Whales and Whaling

Great Southern Regional Development Committee, 1978, Submission to the Inquiry into Whales and Whaling

Gulland, J.A., 1974, The Management of Marine Fisheries, University of Washington Press, Seattle

_____, 1976, 'Antarctic Baleen Whales: History and Prospects', Polar Record, Vol. 18, No. 112, pp.5-13

_____, 1978, Submission to the Inquiry into Whales and Whaling

Hamilton, R.J., M. Long and M.Y. Raie, 1972, 'Sperm Whale Oil. Part 3 Alkanes and Alcohols', American Oil Chem. Soc. J., Vol. 49, pp.307-10

Hardy, A., 1967, Great Waters, Collins, London, UK

Harms, M.G., 1978, 'Jojoba', Submission to the Inquiry into Whales and Whaling

Harrison. R.J. (ed.), 1972, Functional Anatomy of Marine Mammals, Vols. I-III, Academic Press, London, UK

_____, 1978, Discussions with the Inquiry into Whales and Whaling, Cambridge, UK

Heinsohn, G.E., 1978, Submission to the Inquiry into Whales and Whaling

Hilditch, T.P. and P.N. Williams, 1964, Chemical Constitution of Natural Fats, 4th ed., Chapman and Hall Ltd, London, UK

Holloway, P.J., 1968, 'The Chromatographic analysis of Spermaceti', Journal of Pharmacy and Pharmacology, Vol. 20

Holt, S.J., 1978, 'A Review of Current Whale Harvesting Strategies', Report of Inquiry into Whales and Whaling, Vol. 2

_____, and L.M. Talbot 1978, 'New Principles for the Conservation of Wild Living Resources', Wildlife Monograph No. 59, The Wildlife Society

Hutton, B. J., 1978, Submission to the Inquiry into Whales and Whaling

IUCN, 1978, Submission to the Inquiry into Whales and Whaling, International Union for Conservation of Nature and Natural Resources

IWC, 1964, International Whaling Commission Fourteenth Report, London, UK

———, 1969, International Whaling Commission Nineteenth Report, London, UK

———, 1977, International Whaling Commission Twenty-seventh Report, Cambridge, UK

———, 1978, International Whaling Commission Twenty-eighth Report, Cambridge, UK

IWC/FAO, 1969, 'Report of the IWC/FAO Working Group on Sperm Whale Stock Assessment', in IWC (1969), pp.39-83

IWC/29/24, 1977, 'Draft Revision of the Text of the International Convention for the Regulation of Whaling 1946', International Whaling Commission, mimeo.

IWC/SC/29/Rep.5, 1978, 'Humane Killing', in IWC (1978), pp.90-2

IWC/30/4, 1978, Scientific Committee Report, 1978 meeting, mimeo.

IWC/30/4 Annex J, 1978, 'Report of the Subcommittee on Humane Killing Techniques', in IWC/30/4

IWC/30/4 Annex O, 1978, 'Alternative Whale Management Procedures', in IWC/30/4

IWC/SC/30/Rep.5, 1978, 'Humane Killing and Cetacean Intelligence', report to IWC Scientific Committee, Cambridge, UK, mimeo.

Jacobs, M.S., 1972, '"Intelligence" in the Large-Brained Cetacea', paper submitted to US Congress for Hearings on Marine Mammal Protection Act, mimeo.

_____, 1974, 'The Whale Brain: Input and Behaviour', in J. McIntyre (ed.), Mind in the Waters, Charles Scribner's Sons, New York, pp.78-83

_____, 1978, 'Position Statement on Whale Brain and its Intelligence Potential', Submission to the Inquiry into Whales and Whaling

Japan Whaling Association, 1978, 'The Whaling Controversy: Japan's Position and Proposals', Submission to the Inquiry into Whales and Whaling

Jellinek, J.S., 1970, Formulation and Function of Cosmetics, translated from German by G.L. Fenton, John Wiley & Sons Inc.

Jerison, H.J., 1978, 'Brain and Intelligence in Whales', Report of Inquiry into Whales and Whaling, Vol. 2

Kirkwood, G.P., 1978, 'Maximum Likelihood Estimation of Population Sizes Using Catch and Effort Data', paper SC/30/Doc.12 to IWC Scientific Committee, Cambridge, UK, mimeo

_____, K.R. Allen and J.L. Bannister, 1978, 'An Assessment of the Sperm Whale Stock Subject to Western Australian Catching', Submission to the Inquiry into Whales and Whaling

Knott, R.H., 1978, Submission to the Inquiry into Whales and Whaling

Larkin, P.A., 1977, 'An Epitaph for the Concept of Maximum Sustainable Yield', Trans. Am. Fish. Soc., Vol. 106, No. 1, pp.1-11

Laws, R.M., 1962, 'Some effects of Whaling on the Southern Stocks of Baleen Whales', in E.D. Le Cren and M.W. Holdgate (eds), The Exploitation of Natural Animal Populations, Blackwell, Oxford, pp.137-58

Lende, R.A., and S. Akdikmen, 1968, *J. Neurosurg.*, Vol. 29, pp.495-9

Lillie, H.R., 1955, *The Path Through Penguin City*, Ernest Benn, London, UK

Lilly, J.C., 1976, 'The Rights of Cetaceans under Human Laws', *Oceans*, March, pp.67-8

_____, 1977, 'The Cetacean Brain', *Oceans*, July/August, pp.4-6

_____, 1978, *Communication between Man and Dolphin*, Crown Publishers, New York

Lockyer, C., 1972, 'The Age at Sexual Maturity of the Fin Whale (*Balaenoptera physalus*) Using Annual Counts in the Ear Plug', *Conseil Internationale pour l'Exploration de la Mer, Journal*, Vol. 34, No. 2, pp.276-94

McHugh, J.L., 1974, 'The Role and History of the International Whaling Commission', in W.E. Schevill (ed.), *The Whale Problem: A Status Report*, Harvard University Press, Cambridge, USA, pp.305-35

_____, 1976, 'The Whale Problem: A Status Report - A Book Review and Perspective', *Ocean Development and International Law J.*, Vol. 3, No. 4, pp.389-411

Mackintosh, N.A., 1942, 'The Southern Stocks of Whalebone Whales', *Discovery Repts*, 22, pp. 197-300

_____, 1966, 'The Distribution of Southern Blue and Fin Whales', in K.S. Norris (ed.), *Whales, Dolphins and Porpoises*, University of California Press, Berkeley and Los Angeles, USA, pp.125-44

Mercer, M.C., 1978, 'Summary Notes on Canadian Whaling Policy', Canadian Government, mimeo.

Mitchell, E., and M. Stawski, 1978, 'A Bibliography of Whale Killing Techniques, Especially "Humane" Methods', paper SC/30/Doc.38 to IWC Scientific Committee Meeting, Cambridge, UK, mimeo.

Monitor, 1978, Submission to the Inquiry into Whales and Whaling by Monitor, the Conservation, Environmental and Animal Welfare Consortium

Morgane, P.J., and M.S. Jacobs, 1972, 'Comparative Anatomy of the Cetacean Nervous System', in R.J. Harrison (ed.), Functional Anatomy of Marine Mammals, Academic Press, London, Vol. I, pp.117-244

_____, 1974, 'The Whale Brain: The Anatomical Basis of Intelligence', in J.McIntyre (ed.), Mind in the Waters, Charles Scribner's Sons, New York, pp.84-93

_____, 1978(a), 'Whale Brain and Intelligence', Submission to the Inquiry into Whales and Whaling

_____, 1978(b), 'Whale brains and their Meaning for Intelligence', Report of Inquiry into Whales and Whaling, Vol. 2

Muir, E., 1954, An Autobiography, Hogarth Press, London, UK

National Academy of Sciences, 1977, Jojoba: Feasibility for Cultivation on Indian Reservations in the Sonoran Desert Region, National Academy of Sciences, Washington DC, USA

Nieschlag, H.J., G.F. Spencer, R.V. Madrigal and J.A. Rothfus, 1977, 'Synthetic Wax Esters and Diesters from Crambe and Limnanthes Seed Oils' Industrial and Engineering Chemistry Prod. Research and Development, Vol. 16, No. 3, pp.202-7

Nishiwaki, M., 1978, Submission to the Inquiry into Whales and Whaling

Norris, K.S. (ed.), 1966, <u>Whales, Dolphins and Porpoises</u>, University of California Press, Berkeley and Los Angeles, USA

Office of Regional Administration, Albany, 1978, Submission to the Inquiry into Whales and Whaling (on behalf of the Albany Town Council, the Albany Shire Council, the Albany Tourist Bureau, the Albany Port Authority and the Albany Regional Development Committee)

Ohsumi, S., 1977, 'A Preliminary Note on Japanese Records on Death-times for Whales Killed by Whaling Harpoon', in <u>International Whaling Commission Twenty-seventh Report</u>, Cambridge, UK, pp.204-5

Ommanney, F.D., 1971, <u>The Lost Leviathan</u>, Dodd, Mead and Co.

Paterson, R., 1978, Submission to the Inquiry into Whales and Whaling

Payne, R., 1978, Discussions with the Inquiry into Whales and Whaling, New York, USA

Pilleri, G., 1971, 'Intelligence Under Water - the Dolphin Brain', <u>Nautilus</u>, Vol. 9, CIBA-GEIGY Ltd, Basle, Switzerland

———, 1975, <u>Die Geheimnisse der Blinden Delphine</u>, Hallway Verlag, Berne

———, and P.J. Morgane, 1978, 'Large Brains and Intelligence - the Meaning of Special Brain Development Seen in Toothed and Baleen Whales', paper to IWC Scientific Committee Meeting, Cambridge, UK, mimeo

Plummer, F.R., and B.E. Bartlett, 1975, 'Mercury Distribution in Laying Hens Fed Whalemeal Supplement', <u>Bull. of Environ. Contam. and Toxicol.</u>, Vol. 13, No. 3, pp.324-9

Project Jonah, 1978, Submission to the Inquiry into Whales and Whaling

Pryor, K.W., R. Haag and J. O'Reilly, 1969, 'The Creative Porpoise: Training for Novel Behaviour', J. Exper. Anal. of Behaviour, Vol. 12, pp.653-61

_____, 1973, 'Behaviour and Learning in Porpoises and Whales', Naturwissenschaften, Vol. 60, pp.412-20

_____, 1978, Letter to the International Whaling Commission, in IWC/SC/30/Rep. 5

Recchuite, A.D., 1973, 'Case Study in the Development of Sperm Oil Substitutes Part 1, N.L.G.I. Spokesman, November, pp.298-304

Rice, D.W., 1977, 'A List of Marine Mammals of the World' (3rd ed.), National Oceanic and Atmospheric Administration, Washington DC, USA

Ridgway, S.H., 1972, Mammals of the Sea: Biology and Medicine, Charles C. Thomas, Springfield, Illinois, USA

Robinson, E., and P. Waters, 1978, Submission to the Inquiry into Whales and Whaling

Royal Society for the Prevention of Cruelty to Animals, Western Australia, Inc., 1978, 'Methods Used in Taking Whales', Submission to the Inquiry into Whales and Whaling

Sagarin, E. (ed.), 1957, Cosmetics - Science and Technology, John Wiley & Sons, Inc.

Scarff, J.E., 1977, 'The International Management of Whales, Dolphins and Porpoises: An Interdisciplinary Assessment', Ecology Law Quarterly, Vol. 6, pp.323-638

Scheffer, V.B., 1976, 'Exploring the Lives of Whales', National Geographic, Vol. 150, pp.752-66

_____, 1978, 'Are There Special Features of Whales Which May Make Their Conservation Important?', Submission to the Inquiry into Whales and Whaling

Sigma Chemicals Pty Ltd, 1978, 'The Insignificance of Mercury in Whales', Submission to the Inquiry into Whales and Whaling

Singer, P., 1978, 'The Ethics of Whaling', Submission to the Inquiry into Whales and Whaling

Slijper, E.J., 1962, Whales, Hutchinson, London, UK

Smith, W.G. Inc., 1978, Submission to the Inquiry into Whales and Whaling

South Australian Government, 1978, Submission to the Inquiry into Whales and Whaling

Spencer, G.F. and W.H. Tallent, 1973, 'Sperm Whale Oil Analysis by Gas Chromatography and Mass Spectrometry', American Oil Chemists Society Journal, Vol. 50, pp.202-6

Surmon, L.C. and M.F. Ovenden, 1962, 'The Chemistry of Whale Products', Die Suid-Afrikaanse Industriële Chemikus, Vol. 16, pp.62-72

Taverner, M.R., 1975, 'Use of Whale Meal and Whale Solubles as Dietary Protein for Growing Pigs and Their Effects on the Accumulation of Mercury in Tissues', Australian Journal of Experimental and Animal Husbandry, Vol. 15, pp.363-8

Talbot, L.M., 1975, 'Maximum Sustainable Yield: an Obsolete Management Concept', <u>Trans. N. Amer. Wildl. Nat. Res. Conf.</u>, Vol. 40, pp.91-6

_____, 1978, Discussions with the Inquiry into Whales and Whaling, New York, USA

Tranter, D.J., 1978, 'Certainties, Uncertainties and Speculations', Submission to the Inquiry into Whales and Whaling

US Congress, HR 3465, 1975, <u>Whaling,</u> Whale Oil and Scrimshaw, Hearings before the Subcommittee on Fisheries and Wildlife Conservation and the Environment of the Committee on Merchant Marine and Fisheries, House of Representatives, Ninety Fourth Congress. First session on HR 3465 'To Release Sperm Oil from National Stockpiles', 9 and 10 June, 1975, US Govt. Printing Office, Washington DC, USA

US Delegation to the International Whaling Commission Meeting, 1978, Discussions with the Inquiry into Whales and Whaling

US Public Law 92-522: Marine Mammal Protection Act 1972

US Public Law 94-359: Endangered Species Act 1973

Walker, E.P., 1975, <u>Mammals of the World</u> (3rd ed.), John Hopkins University Press, Baltimore and London

Warshall, P., 1974, 'The Way of **Whales**', in J. McIntyre (ed.), <u>Mind in the Waters</u>, Charles Scribner's Sons, New York

Warth, A.H., 1947, <u>The Chemistry and Technology of Waxes</u>, Reinhold

Watkins, W.A., 1977, 'Acoustic Behaviour of Sperm Whales', <u>Oceanus</u>, Vol. 20 No. 2, pp.50-8

Western Australian Government, 1978, Submission to the Inquiry into Whales and Whaling

Wildlife Preservation Society of Australia, 1978, Submission to the Inquiry into Whales and Whaling

Wolman, A.A., and D.W. Rice, 1978, 'Current Status of the Gray Whale', paper SC/30/Doc.34 to IWC Scientific Committee Meeting, Cambridge, UK, mimeo.

World Wildlife Fund, 1978, Submission to the Inquiry into Whales and Whaling

RESOURCES FROM FRIENDS OF THE EARTH

Soft Energy Paths
Toward a Durable Peace

By Amory B. Lovins, introduction by Barbara Ward

Soft Energy Paths crowns the critical work of Amory Lovins and FOE with a new program for energy sanity. This is an epoch-making book, the most important Friends of the Earth has published.

Lovin's soft path can take us around nuclear power, free industrial nations of dependence on unreliable sources of oil and the need to seek it in fragile environments, and at the same time enable modern societies to grow without damaging the earth or making their people less free.

Soft Energy Paths provides a conceptual and technical basis for more efficient energy use, the application of appropriate alternative technologies, and the clean and careful use of fossil fuels while soft technologies are put in place.

Harper & Row/Colophon edition
231 pages
$3.95

Pathway to Energy Sufficiency
The 2050 Study

By John Steinhart

Pathway to Energy Sufficiency is a hardheaded, hopeful look at an American society seventy years in the future. The authors draw a picture of what it would be like to live along a soft energy path, using 36 percent of the energy we used in 1975. They tell how this would affect our houses, travel, countryside, factories, gardens, and public institutions. They find this low energy society decentralized, more rural, less hurried, and healthier.

"The benefits that will accrue to a society that accepts the low-energy scenario could be most attractive. If people want a world in which people restrain their numbers and appetites, people can achieve it—and will prefer it to the grim alternatives."
David R. Brower

96 pages, illustrated
$4.95

Frozen Fire:
Where Will It Happen Next?
By Lee Niedringhaus Davis

An LNG accident could be as bad as a reactor meltdown, and the major exporters of liquefied natural gas are OPEC countries.

But the gas industry wants to make it common, building terminals on most American coasts, Europe and Japan, bringing 125,000-cubic-meter shiploads of it in from the Middle East, Indonesia, and Alaska.

US trade in LNG is insignificant now, and world trade is small compared to industry plans. So there is still time to avert the dangers. If we do not, recent horrors with liquefied gases in Spain, Mexico, England, Abu Dhabi, and Staten Island may be the harbingers of a fearful future.

Lee Davis's book is the first major study of LNG prepared for the general reader. It tells everything you should know about the stuff before they try to put it in your town.

"The scope and logic of *Frozen Fire* should make it the Bible of anti-LNG groups throughout the world. It could even make converts of government agencies and the gas industry, preventing the inevitable LNG holocaust."
—*Gene and Edwina Cosgriff*
B.L.A.S.T.

"An impressive and powerful compendium of information on the dangers of another chemical threat to the so-called civilized world."
—*John G. Fuller*

256 pages
Cloth: $12.50
Paperback: $6.95

The Whale Manual
By Friends of the Earth Staff

The Whale Manual lays out the latest facts and figures about the great whales. Population estimates, habitats, and how quickly they are being killed, by whom, for what, and how—and what could be used as alternatives to whale products.

A special section outlines Friends of the Earth's controversial program to preserve the endangered Bowhead, with the help of the Eskimos who hunt it.

The Whale Manual is an invaluable source for readers committed to saving our planet's largest creatures.

168 pages
$4.95

To: Friends of the Earth
124 Spear Street
San Francisco, CA 94105

Please Join Us.

☐ Please enroll me for one year in the category checked, entitling me to *Not Man Apart* and discounts on selected FOE books.
(*Contributions to FOE are not tax-deductible.*)

☐ Regular = $25 ☐ Spouse = add $5
☐ Supporting = $35* ☐ Life = $1000***
☐ Contributing = $60** ☐ Patron = $5000***
☐ Sponsor = $100** ☐ Retired = $12
☐ Sustaining = $250** ☐ Student/Low Income = $12

*Will receive free a paperback volume from our *Celebrating the Earth* Series.
**Will receive free a volume from our *Earth's Wild Places* Series.
***Will receive free a copy of *Headlands* (our award-winning, gallery-format book).
☐ Check here if you do not wish to receive your bonus book.

☐ Please accept my *deductible* contribution of $ _____ to Friends of the Earth Foundation (*checks must be made to FOE Foundation*).

Please send me the following FOE Books:

Number	Title, price (*members' price*)	Cost
_____	*Frozen Fire* at $12.50 ($10.00)	_____
_____	paperback *Frozen Fire* at $6.95 ($5.95)	_____
_____	*Energy Controversy* at $12.50 ($10.00)	_____
_____	paperback *Energy Controversy* at $6.95 ($5.95)	_____
_____	*Progress As If Survival Mattered* at $6.95 (5.75)	_____
_____	*SUN! A Handbook for the Solar Decade,* at $2.95 ($2.25)	_____
_____	*Soft Energy Paths,* H&R ed. at $3.95 ($3.25)	_____
_____	*The Whale Manual* at $4.95 ($3.95)	_____
	Other FOE Titles:	

_____
_____
_____

Subtotal _____
6% tax on Calif. delivery _____
Plus 5% for shipping/handling _____

☐ Send full FOE Books catalogue. TOTAL _____
☐ VISA ☐ Mastercharge
Number _____ Expiration date _____
Signature _____
Name _____
Address _____
City _____ State _____ Zip _____